Estimating for Builders and Qu

BH NEWNES

Newnes
An imprint of Butterworth-Heinemann Ltd
Halley Court, Jordan Hill, Oxford OX2 8EJ

 PART OF REED INTERNATIONAL BOOKS

OXFORD   LONDON   GUILDFORD   BOSTON
MUNICH   NEW DELHI   SINGAPORE   SYDNEY
TOKYO   TORONTO   WELLINGTON

First published 1991

British Library Cataloguing in Publication Data
Buchan, R. D.
  Estimating for builders and quantity surveyors.
  1. Buildings. Construction. Estimating
  I. Title   II. Fleming, F. W.   III. Kelly, J. R.
  692.5

ISBN 0 7506 0041 1

Filmset by Key Graphics, Aldermaston, Berks
Printed and bound in Great Britain by
M & A Thomson Litho Ltd., East Kilbride, Scotland

Contents

Preface

This book has been written as an introduction to estimating for students taking courses in building and surveying at colleges, polytechnics and universities up to BSc level.

The content is devoted to the calculation of rates for the items in bills of quantities. All examples are based on items measured in accordance with the seventh edition of the *Standard Method of Measurement of Building Works* (SMM7). The chapters generally cover one topic or one trade; reference to the relevant sections of SMM7 is made in the chapter headings. The examples have been chosen to reflect both traditional and up-to-date technology.

A chapter has been included on computerized estimating, and the appendix shows how a spreadsheet program can be developed for the calculation of the cost of mortars, renders and screeds. The principles illustrated can be applied to any form of calculation.

The examples in the book have been prepared using a spreadsheet program which has resulted in the rounding off of subtotals and totals, so that in one or two instances there would appear to be an error of a penny or two. This does not affect the final rate calculated, as the computer always works to a minimum of eight places of decimals (although perhaps only displaying or printing the result to two places).

A floppy disk with the spreadsheet from the appendix and some of the worked examples from other chapters is available from the authors. An order form is provided at the end of the book.

Chapters on tendering and tendering strategy are included with the objective of setting estimating, as distinct from tendering, in its proper context.

The policy for all the trades included has been to provide a core of the most useful examples and to reinforce areas which appear to have been less well explored by other authors by giving rather fuller explanation. Feedback, via the publishers, on areas which the reader may feel require more detailed explanation would be most welcome.

Finally, we have had much help from many colleagues in the construction industry – too many, in fact, to name individually here. To all of them we are most grateful and trust that a general acknowledgement of that help here will be acceptable.

R. D. Buchan
F. W. Fleming
J. R. Kelly

Note

The cost of labour in the examples is given as:

- Tradesman   £6.10 per hour
- Labourer   £5.30 per hour

Both rates are inclusive of tool and clothing allowances. However, where a labourer is paid an additional sum for continuous skill or responsibility, this sum has been added net to the rate. (See the examples in Chapter 2.)

Introduction

Estimating is the activity of pricing work in preparation for making an offer to carry out that work for a sum of money. For building work the offer is made by a builder to a client who wishes building work to be carried out.

The tender sum offered will be made in the context of a form of building contract which will describe the conditions under which the sum offered may be varied. At the time of making the offer the builder knows that his pricing must be accurate, since the standard forms of building contract do not permit the contractor to recover additional money from the client as a result of an error in estimating.

The chapters in this book describe the method used by the estimator to build up prices or rates for items described in SMM7 format. Also included are chapters on computer aided estimating and on tender strategy.

Each chapter is a self-contained unit and, notwithstanding the fact that prices in the chapters interrelate, chapters can be read singly and in any order.

The invitation to tender

The invitation to tender given by a client to the builder may arise in different ways. For simple or low value work, such as domestic decorating, a small house extension or even the changing of all fluorescent tubes in an office, the client will most probably invite the builder to the premises, describe the work to be carried out and wait for the offer. The builder's estimator will measure the work, price it and make an offer based upon the builder's own terms and conditions. The offer may also contain details of specification but the builder, in this situation, will be bound to carry out the work based upon the normal trade practice of the time.

For work of a more detailed nature, for instance the construction of a house or a small light industrial unit, the client will traditionally approach an architect, who will complete the design work and submit the drawings to the builder together with a specification and details of a standard form of contract. The builder will measure the quantities of work, price it and make an offer based upon the drawings, the specification and the standard form of contract selected by the architect.

As an alternative the client may approach a contractor and request that the contractor make an offer for the complete package of design and build. Once reserved for the type of work described in the previous paragraph, this method of tendering is becoming more common and the scale of projects is becoming more complex. A procedure for the selection of a design/build contractor is described later in this chapter.

Finally, for work which is more complex than any of the above – which, measured in terms of value, will cover the majority of building work – the client will traditionally approach an architect, who will receive the statement of the client's requirements and carry out initial design work, usually a sketch. The design is passed to a quantity surveyor who will make a prediction of the likely tender value. If this is acceptable to the client the architect will proceed with the detailed design in conjunction with structural, mechanical and electrical engineers and other consultants. The detailed design is given to the quantity surveyor who will measure all items of work in accordance with a set of rules for measurement and prepare a book, called the bill of quantities, which contains the specification and the measured items of work.

The set of rules for measurement is compiled by agreement between the Royal Institution of Chartered Surveyors (RICS) and the Building Employers' Confederation (BEC). This set of rules, termed the *Standard Method of Measurement for Building Works*, was first compiled in 1922 and is currently in its seventh edition (SMM7).

Scottish practice used the *Scottish Mode of Measurement* until 1963, when SMM5 was adopted for use in Scotland. Since that time both SMM6 and now SMM7 have been accepted for use in Scotland.

SMM7 is important to the builder's estimator as it describes within the coverage rules those items of work which are deemed to be included with main items of construction. For example, in the groundwork rules, D20.6, 'working space allowance to excavations' is measured in square metres and is deemed to include the cost of work associated with 'additional earthwork support, disposal, backfilling, work below ground water level and breaking out'. An estimator who did not realize the extent of the 'deemed to be included' items might underprice the item.

The bill of quantities, prepared in accordance with SMM7, is sent with the drawings to the builder. The builder's estimator will price the items described and will include with the pricing of the item those aspects which have been defined by SMM7 as 'deemed to be included'.

The invitation to tender will be expressed in one of the four ways described above, that is as a request to carry out work for a sum based upon: a verbal or written outline description of what is required; or drawings, specification and a form of contract; or drawings, bill of quantities and a form of contract; or finally the provision of a full design and build service.

The estimating method

The builder's estimator, having received the necessary instructions and documentation from the client, will set about providing the estimate. Initially the estimator must break the work down into small segments to be priced. If the estimator is provided with a bill of quantities this will have been done by the quantity surveyor; otherwise the estimator must prepare a similar document for pricing.

The estimator will identify packages of work which will be carried out by subcontractors. For each work package the estimator will select a number of subcontractors who are willing to tender and will send to each a copy of the relevant section of the bill of quantities. The successful tenderer's prices will, with an addition for the builder's overheads and profit, be incorporated into the builder's offer.

For the remaining work the estimator will build up prices in the manner which is described in the chapters of this book. A rate for an item of work comprises the cost of the resources plus an allowance for overheads and profit. The resources comprise labour, material and plant.

For all work the estimator will have a table of labour constants, i.e. the time that it takes one or more operatives to carry out a task. The constant is normally expressed as an output of work per hour. By totalling the hours and multiplying by a cost of labour per hour the estimator can determine the labour cost associated with the work item.

Plant is priced on the same basis as labour if the plant is used in a task; for example, an excavator digging a drain trench will be priced on the basis of the amount of work that it can do in one hour. Static plant such as scaffolding, site hutting or even a tower crane will be priced on the basis of the time that it is required on site. The cost is added to the tender as a lump sum and is not priced per unit of task, e.g. £1.50 per m^3.

Material is priced at its purchase price plus allowances for delivery, off-loading, storage and placing in position. In addition an allowance must be made to cover wastage which may arise as follows:

Handling breakages Brittle materials such as bricks, precast concrete paving slabs etc. are prone to breakage even with the advances in mechanized handling of material which is wrapped and delivered on pallets.
Site losses Loose material such as sand and aggregate delivered and tipped on site will to some extent be trampled into the ground, washed away by rain etc. and an allowance has to be made for this.
Cutting losses Material delivered in sheet form such as plasterboard, plywood, glass, carpet etc. is cut to fit. Where the material is patterned or has to be cut to a difficult shape, losses can be considerable.

Overheads and profit are added to all work items to cover the cost of running the builder's establishment and to provide a profit. The builder's establishment cost comprises the cost of running the head office, which includes such departments as estimating, buying, planning and accounts, together with those departments which are essential to any commercial concern such as marketing, strategic planning, public relations, personnel and training.

The majority of estimators will price their rates without overheads and profit, leaving the amount of this addition to be decided by the board of directors. The builder has four choices for the distribution of overheads and profit:

1 To add a percentage to all individual rates. This has an advantage where a variation to the work adds additional quantity, since the builder will recover additional overheads and profit, but the converse is also true.
2 To add the total amount to one item in the preliminaries section, usually to the supervision item. This has the advantage that the amount cannot be reduced, and puts the builder in a strong negotiating position where a claim is made for additional supervision owing to the contract period being lengthened.
3 To reveal the overheads and profit addition by making an addition to the summary total at the end of the bills of quantities. This blanket addition has a disadvantage if an overestimate of provisional sums and provisional quantities is made in the bills of quantities. A remeasurement of these items may lead to a reduction in the contract sum and hence also in the blanket addition.
4 To add the amount selectively to specific items in the bills of quantities to improve cash flow. This has advantages, but is also dangerous where work is subject to remeasure.

The reasons considered by the builder for the addition of overheads and profit are discussed at length in Chapter 22 on tender strategy. In this book a 20 per cent addition is used for the purposes of illustration.

Tender procedure

From the client's viewpoint there are three ways in which to select a builder to carry out work. These are as follows.

Open tendering

Under this method an advertisement is placed in a trade journal stating that the client wishes certain work to be carried out and inviting builders to apply for the tender documents. Any builder, irrespective of size or capability, may apply for the documents and tender. Usually a deposit is required in exchange for the documentation; the deposit is refunded upon receipt of a bona fide tender.

The disadvantage of this method is in the selection of a contractor of whom the client can be confident. The method was disapproved of in the Banwell Report of 1964 and is not greatly used today. It is interesting to note that in the USA open tendering is used for all federal building work and the lowest tenderer will, except in unusual circumstances, be awarded the contract. The only provision made by the federal authorities is that the builder must be 'bonded'; that is, the builder must obtain a bond from a bank or an insurance company in which the latter guarantees that in the event of the builder failing to carry out the work the federal authority will be reimbursed for the full cost of having the work completed by another builder.

Selective tendering

Under this method the client or the architect and quantity surveyor compile a list of the preferred builders for the project. Those on the list of builders will have been chosen for their established skill, reliability and proven competence for the type of work envisaged.

Local authorities and government agencies keep a general list to which builders can apply to join. Builders are carefully vetted before being put on the general list. It is from this file of approved builders that a list is made of those who will be invited to tender for a specific project.

The Code of Procedure for Single Stage Selective Tendering was produced by the National Joint Consultative Committee for Building (NJCC) and published by RIBA Publications in a revised format in 1989. It makes the recommendation that no more than six builders should be selected for any tender. Many commentators regard the maximum number stated as too high, bearing in mind the high cost of tendering.

Negotiated tender

Under this method only one builder is approached, normally because the skills of the builder are such that the architect and the other members of the design team wish to take advantage of the builder's specialist knowledge at the design stage.

Following the completion of the design the builder will price the bill of quantities and then enter into a negotiation with the quantity surveyor. For this type of situation the NJCC has issued the Code of Procedure for Two Stage Selective Tendering, which is discussed later.

Single stage selective tendering

Invitation

The Code of Procedure for Single Stage Selective Tendering makes the recommendation that the first approach to a builder is a preliminary enquiry. The preliminary enquiry should indicate that a list of tenderers is being drawn up for a specific project and ask whether the builder wishes to be considered for inclusion in that list. The enquiry should indicate that the builder's inability to accept this invitation will not prejudice a future invitation. This is important since many builders think that it is better to submit a high 'cover' price than decline an invitation. In fact, the object of the design team is to compile a list of builders who wish to submit a serious tender.

Decision to tender

Having received the invitation the builder must make a decision as to whether to tender or not. This decision will be based upon the policies formed in the company plan and the determination of whether the likely cash flow demand of the project is conducive to the company cash flow and borrowing arrangements. These factors are discussed at length in Chapter 22 on tender strategy. In addition, the builder must consider the resource demand of the proposed project in comparison with the resources which the company has available, and also whether these resources can easily be deployed in the geographic area of the proposed site.

A further factor to be considered is the type and form of contract. Currently there are many ways of procuring buildings and the builder must consider whether it is desirable to operate under a form of contractual arrangement which is not familiar. The builder must also ascertain the extent of risk associated with a particular project and decide whether the risk is worth taking.

The final factor to ascertain is the extent of the competition, since this will affect the probability of winning the tender; for example, a builder may make a decision not to tender in a competition with more than four others. Although not stated in the Code of Procedure for Single Stage Selective Tendering, the design team should indicate in the preliminary enquiry the target number of builders to be included in the tender list.

The tender

On the date stated in the preliminary enquiry the documents forming the tender pack are dispatched to those builders who have agreed to tender and have been selected for the final tender list. A standard letter should go to those who indicated their willingness to tender but have not been selected for the final list, stating that they will be invited again in the future. The tender pack contains: plans, elevations and sections, schedules and component drawings; the bill of quantities; the tender form (see Figure 1.1); an envelope for the return of the tender form marked with a code which indicates to the quantity surveyor the name of the tenderer; and an envelope for the return of the priced bill of quantities, similarly coded.

The estimator begins the preparation of the tender by dividing the work into that which the builder will undertake and that which will be subcontracted. A copy of the bill of quantities is passed to the buying department so that quotes for materials may be obtained. It is at this point that most builders discover who the competition are, simply by selecting the easiest defined or least common material in the bill of quantities and asking the supplier if any other builder has requested an identical quantity of the same material. It is very difficult to keep a tender list confidential to the design team.

Tradition states that four weeks is sufficient for any tender. In practice this is not so, and the design team should plan for at least six weeks where the builder, or one of his subcontractors, is expected to carry out design work in accordance with a performance specification in the bills of quantities.

The estimator will generally price all items in the bill of quantities at cost with no mark-up for overheads and profit. Following this exercise the management of the builder will decide upon the overheads and profit percentage mark-up. They will take account of the items of risk in the

```
Tender for:   (description of the works)

To:   (Client)

Sirs,

We having read the conditions of contract and bills of
quantities delivered to us and having examined the drawings
referred to therein do hereby offer to execute and complete
in accordance with the conditions of contract the whole of
the works described for the sum of:

     £ ..........................

and within .... weeks from the date of possession.

We agree that should obvious errors in arithmetic be
discovered before acceptance of this offer in the priced
bills of quantities submitted by us these errors will be
dealt with in accordance with Alternative 2 contained in
Section 6 of the Code of Procedure for Single Stage
Selective Tendering.

This tender remains open for acceptance for .... weeks from
the date fixed for the submission or lodgement of tenders*.

Dated this ........ day of ........ 19  ..

Name:

Address:

Signature:

Witness 1:

Witness 2:

* Scotland only: 'unless previously withdrawn'.
```

Figure 1.1   *Tender form*

project within the percentage and will adjust the projected cash flow by selective addition of the overall amount of overheads and profit (see Chapter 22 on tender strategy).

If a particular risk is so great that the builder is unable to account for it within a realistic tender, the tender document may be qualified. For example, an item in a bill of quantities may state (contrary to SMM7): '17C   Carry out all necessary repairs to the Welsh slate roof – item.' In this situation a lump sum is being requested. The builder may consider the risk of pricing this item as a lump sum to be too great, and therefore inserts in the tender document: 'We offer to carry out the work ... for the lump sum of £300,000, which includes a provisional sum of £1000 for repairs to the roof in the place of bill of quantities item 17C.' Although such a statement is a legally valid offer under the law of contract, the Code states that in such circumstances the tenderer should be given the opportunity of withdrawing the qualification. If the builder is unwilling to withdraw the qualification, the tender should be rejected.

The logic behind this argument is that tenderers have the opportunity to bring such matters to the attention of the design team during the tender period, and a correction, where this is thought appropriate, can be issued to all tenderers. In practice many tenders received are qualified, and the design team are put in a difficult position when under pressure from the client to accept the lowest.

Receipt of tenders

Two days before the date for the receipt of tenders the quantity surveyor should determine which tenders have not been received by reference to the coded envelopes. The builders whose tenders have not been received should be telephoned to ascertain whether a tender is to be submitted. In the case where tenders have been delayed in the post this permits another copy to be sent by post, courier or fax.

As soon as possible after the time for the receipt, all tenders should be opened. In Scotland the priced bills of quantities will have been sent with the tender in a separate envelope and the bills of the lowest tenderer will be opened for checking. Elsewhere in the UK the priced bills will be requested from the lowest tenderer.

In the event that no arithmetic errors are found and the offer is acceptable to the client, the offer will be formally accepted and a contract entered into. The bills of the remaining tenderers in Scotland will be returned unopened.

The Code of Procedure recommends that immediately after the opening of tenders all but the three lowest should be informed that their tender was unsuccessful. If an arithmetic error is found during the checking of the bills of quantities then two alternatives are given in the Code.

Alternative 1 states that the tenderer should be notified of the error and given the option of confirming the tender or withdrawing. Alternative 2 states that the tenderer should be notified of the error and given the option of confirming the tender or correcting the error. In reality the difficult decision only applies where the correction of the tender puts the tender price higher than that originally. In the case that the tenderer withdraws (alternative 1) or that the corrected tender is higher than the second lowest (alternative 2), the priced bills of quantities of the second lowest are examined and the whole procedure starts again.

Feedback

It is recommended that each tenderer is given a list of the tender sums received so that each can observe their position in the list and a decision can be made with regard to the strategy to be adopted for future tenders. In Scotland, as a separate list to the tenders, each tenderer will receive an alphabetical list of the builders tendering.

Serial tendering

A variation on the above procedure is that of serial tendering in which the builder tenders for a series of similar contracts. The method was popular during the period of industrialized building, particularly in conjunction with school buildings, where a learning curve was anticipated. The rationale was that the builder might accept a small loss on the first school building, but by the time the third had been constructed the builder's operatives would be fully conversant with the system. Serial tendering also guarantees work for an extended period. The method is based on the tenderer pricing a master bill. This will contain the quantities for one school but may include, for instance, three options for the construction of a roof, so that three sets of roofs for one school will appear in the master bill. Similarly the foundations may be measured as both pads and piles (assuming a light steel frame construction).

The tender procedure will be exactly as the above but the builder will know that the pricing of the master bill effectively represents a tender for five schools to be constructed over the following five years. The schools are designed and bills of quantities prepared and priced at the rates in the master bill. Normally the serial contract is based upon the rates being adjusted in accordance with a price index, for example the NEDO index.

Two stage selective tendering

The majority of negotiated tenders, as stated above, arise from a desire to involve the builder in the design phase of a project. The Code of Procedure for Two Stage Selective Tendering describes a method which allows the negotiation of a tender with an element of competition.

First, suitable builders are selected in the same manner as for single stage selective tendering, and a list is compiled of those willing to take part in the initial competitive stage. The tender documents sent to the tendering contractors comprise outline sketches of the scheme, a summary of the client's requirements and a notional bill of quantities. The notional bill of quantities is a document containing an outline specification and provisional item descriptions and quantities.

The Code states that tenderers should be given five weeks to prepare their tender. It recommends that the tender is submitted under the same rules as for single stage selective tendering, with the same procedures for notification, errors etc.

Many design teams, however, do not use the preliminary tender based upon the notional bill as the sole method of selection. Instead they will invite each tendering builder to a meeting to discuss the builder's approach to the project. The approach will be discussed in terms of the builder's contribution to the design, the previous experience of such schemes, the personnel to work with the design team and their previous experience. The aim is to ascertain whether the builder and the design team can work well together. It is the combination of these meetings and the preliminary tenders which determines the successful builder.

An agreement will be made with the successful builder regarding the cost of the builder's input before the commencement of the design stage. The design stage is undertaken. Bills of quantities are prepared by the quantity surveyor and where possible priced at the rates in the notional bill. Those items which cannot be priced on this basis are priced by the builder and are subject to negotiation.

Finally following a successful negotiation the builder will make a formal offer to carry out the work for the sum of money negotiated, and this is formally accepted by the client. A standard form of contract is completed.

This method of tendering allows the choice of a builder who is capable and keen to contribute to the design while retaining the important element of competition.

Selective tendering for design and build

The NJCC published in October 1985 a Code of Procedure for Selective Tendering for Design and Build. Similar to the two stage method described above, this outlines a tender method which is considered fair to all parties. In design/build particularly, a tendering builder is only prepared to pay for the development of the design and tender if there is a good probability of winning the work. The Code's recommendations are as follows.

First, six to twelve selected builders are circulated with a letter of preliminary enquiry, which sets down the nature of the project, the anticipated start and completion dates etc. and asks whether the builder would be interested in tendering for the project.

Those builders who reply positively to the preliminary enquiry are sent a copy of the employer's (client's) requirements. The documents which represent the employer's requirements may be as little as a description of the accommodation required, or may be anything up to a full scheme design undertaken by a full design team employed by the client. However the employer's requirements are expressed, it is anticipated that they will be fully considered at the tender stage and will not be subsequently changed.

Tendering builders are called for an interview with the employer and his advisers. At the interview the following will be discussed:

● The form of construction anticipated by the builder
● The builder's organization and anticipated programme of work
● The nature and extent of the design liability insurance.

Following the interview a shortlist not exceeding four builders will be drawn up and formal tender documents will be dispatched.

The Code describes a large number of variations in tender practice which are acceptable at this point. These vary from a straight single design and lump sum tender to multiple alternative submissions and two stage selection. In the latter case the builder may submit alternative schemes and give rates for major items of work. Following an analysis of the schemes received, one builder may be given the go-ahead to develop particular schemes further. The resulting tender may or may not be open to negotiation based upon the rates originally submitted.

The final scheme prepared by the builder and submitted with the tender is termed the builder's proposals. The builder's proposals are an answer to the employer's requirements, and comprise

● Plans, elevations, sections or typical details
● Information about the structural design
● Layout drawings showing the services to be included
● Specifications of workmanship and material.

If the builder's proposals and tender are acceptable to the employer or client, a contract is signed between the builder and employer based upon the submitted data.

Finally, it is recommended that feedback be given to the unsuccessful, though this is unlikely to be as useful as a simple list of tender prices received.

Under the design/build contract the estimator has responsibility for the measurement and pricing of the work since no bills of quantities are provided. Many design/build builders will engage professional architects and quantity surveyors to carry out their traditional work.

The design/build contract is a lump sum contract payable in stages or by valuation in accordance with the conditions. As there is no provision for a bill of quantities and no schedule of rates, the contract makes provision for a 'contract sum analysis' to permit the valuation of employer's changes and for formula price adjustment if this is included as a contract provision.

There is no provision in the contract for prime cost sums and these should not be included. There is, however, provision for the expenditure of provisional sums included by the employer, and the contract sum analysis should be sufficient to allow this.

2

The cost of labour

The cost of labour represents a high proportion of today's building costs. As a percentage of a bill rate (other than labour-only rates) the figure varies from about 30 to 75 per cent over the various trades. In the overall cost of the building, the cost of labour will account for something like 40 to 60 per cent.

The all-in hourly rate

The complete cost of labour is calculated as an hourly figure which is widely known as the 'all-in hourly rate'. This should not be confused with the term 'all-in rate', which is sometimes used instead of 'bill rate' to describe the price attached to an item in a bill of quantities and which includes labour, material, plant, profit and overhead costs.

There are basically two ways to calculate the all-in hourly rate. The first is to consider what a particular class of operative will cost the building employer for one year and to divide this by the total hours worked in that year. The second is to consider what the same group will cost for one week and to divide by the hours worked that week.

The former is widely used by various general sources of bill rates such as pricing books, computerized libraries of rates and some building magazines. The all-in hourly rate produced can only be a typical average rate, as the producer of the data has no contract in mind and therefore must assume how much, if any, overtime is being worked, the ratio of tradesmen to labourers in squads, the amount of supervision and whether this is done by working or non-productive foremen, and so on. There is no notion of distance from site, so averages can only be given for things like travelling time, fares, lodgings, transport etc.

The all-in hourly rate based on a week's work is produced by estimators for particular circumstances. In this instance the estimator knows how large the contract will be, how far away it is and what work is required to be done, and therefore the resources necessary in terms of operatives, supervision, overtime, travelling etc. In other words a lot of averaging and guesswork becomes a matter of judgement based on the employer's records of previous contracts.

The reader is recommended to study the all-in hourly rates given in the building magazines and price books. Notice the comments regarding the allowances for overtime, bonuses etc.

The Working Rules

Part of the cost of labour is naturally the actual wages paid to the workers carrying out the productive work. The workers are generally referred to as operatives, and we will return to the idea of productive work later in the chapter. The remainder of the cost of labour is made up of a number of payments which the builder must make and which are a direct result of employing operatives. These payments are not always made to the operative and are not always on his behalf.

The following is a classification of labour costs:

- Payments made as a result of negotiation between unions and employers, including hourly wage rates, holidays with pay and overtime rates
- Expenses such as fares, meals and lodgings
- Statutory payments such as national health insurance, employer's liability insurance, sick pay and severance pay
- Payment made as a matter of good business management, including third party liability insurance and adequate supervision.

There are several discrete bodies in the construction industry which agree rates of pay and conditions for different types of operative. The organization whose rules are reviewed in this chapter is the National Joint Council for the Building Industry (NJCBI). There are separate organizations negotiating on behalf of tradesmen such as plumbers and electricians, as well as alternative organizations negotiating for bricklayers, carpenters etc.

The NJCBI comprises members from the following organizations:

- Building Employers' Confederation (comprising regional federations)
- National Federation of Roofing Contractors
- Union of Construction, Allied Trades, and Technicians
- Transport and General Workers' Union
- General, Municipal, Boilermakers' and Allied Trades Union
- Furniture, Timber and Allied Trades Union.

Negotiated hourly rates and the rules of payment of a long list of benefits for the employee are given in the Working Rules, a document produced by the NJCBI. The rates and rules are published with regional amendments through the individual regional offices. In practice, the building employer will use the rules appropriate to the area in which the work is being carried out. The various regional editions of the rules are obtainable from the regional offices of the unions or employers' federations.

Not all building companies pay NJCBI or any other nationally negotiated rates of pay. Negotiated rules are simply standard forms of employment contract. A building contractor may employ labour on any mutually agreed basis as long as statutory requirements are fulfilled. However, no matter what the basis for the employment contract, much the same types of cost are involved. Therefore the job of the estimator will follow similar patterns to those described in the text and shown in the examples.

The examples given in this chapter are based on the Working Rules approved by the Scottish Regional Committee and the NJCBI, August 1988 edition with amendments to July 1989. As the Working Rules are under continuous review, the reader is reminded here that he or she should purchase a copy of the most recent edition and keep it up to date as amendments are published. Amendments are widely circulated in the building journals and the house journals of various professional bodies such as the CIOB and the RICS.

Estimating the cost of labour

It is important that the estimator gets the cost of labour correct; it is a major component of practically every bill rate. Having said that, it is equally important not to be bogged down in a plethora of unwarranted detail in the calculation of that cost.

With the word 'detail' in mind, let us look first of all at the list of factors which could possibly be taken into consideration in calculating the cost of labour. The factors are listed in Table 2.1, and those used in the examples at the end of this chapter are marked with an asterisk.

There are a number of ways in which these items can be costed; these costing options are listed in Table 2.2. For every factor listed in Table 2.1 there is a head office and site overhead/oncost for administering wages, holidays with pay, redundancy schemes, bonus,

Table 2.1   *Factors affecting the cost of labour*

Factor	Used in examples	Costing option (Table 2.2)	Working Rule (WR)
Earnings:			
Hourly rates of pay	*	1	2
Guaranteed minimum weekly earnings	*	1	1
Guaranteed minimum bonus	*	1	1.6
Overtime	*	1	7
Shift work and night work	Not included		8, 9
Distance to site:			
Travelling time	*	1	14
Reimbursement of fares	*	2	14
Provision of transport		2	14
Lodging allowance	*	2	15
Extra payments:			
Continuous extra skill or responsibility	*	1	3
Intermittent responsibility	*	1	4
Holidays:			
Holidays with pay scheme	*	2	10
Public holidays	3.43%	3	11
Retirement and death benefit schemes	*	2	19
Employer's contributions for national insurance	Statutory amounts	3	
Injury and sickness:			
Injury payments	*	3 or 6	16
Sick pay	£1.10/week		16
Payment for work in difficult conditions		1	17
Tool and clothing allowances	*	1	18
Employer's liability insurance	2%	3	
Construction Industry Training Board levy	*	1 or 3	
Redundancy funding	1.5%	3	
Supervision: foremen, chargehands, gangers	By calculation	2	
Tea/meal breaks	4.26%	3	6
Bonus, incentive, productivity schemes	Not included		
Excess overtime allowance	*	3	

* Items so marked are the subject of rates or fixed amounts from the Working Rules and these have been applied in the examples.

Table 2.2 *Costing options*

1   Use the correct hourly rate, allowance or addition together with the correct hours.
2   Add the actual cost to the cost of labour.
3   Add an appropriate percentage to the cost of labour.
4   Reduce the output per man by an appropriate amount.
5   Include the cost in the preliminaries.
6   Include the cost in overheads/oncost.

expenses etc. Costing option 6 in Table 2.2 should therefore be assumed to be in addition to those already listed.

In addition to the factors in Table 2.1 there are a great many items listed in the Working Rules on the subject of employer/employee legislation. These are not easily categorized as part of the cost of labour, but nonetheless they do cost the builder a considerable amount of money which must be recouped in the ordinary course of his business. The following list is by no means exhaustive:

● Liaison with safety representatives appointed by trade unions
● Time off for training safety representatives
● Safety committees with union and employer representation
● Time off and facilities for site, job or shop stewards to carry out their duties (the facilities are generally provided by the builder).

The costs of these are generally accounted for by an addition in the overheads/oncost (option 6 of Table 2.2).

It is interesting to note that options 4 and 5 in Table 2.2 are not thought suitable for any of the items listed in Table 2.1. For example, tea/meal breaks is shown in Table 2.1 as being most suitable for the use of costing option 3. Option 1 does not fit the circumstances, there being no hourly rate, allowance or addition. Option 2 assumes a fixed sum which again is inappropriate. Option 3 is the most feasible in this instance. The loss is easily assessed, occurs on every job and affects every employee. Option 4 would involve a very complicated calculation concerning the proportion of basic hours, productive hours, overtime hours etc., and the result would be no more accurate than that obtained with option 3. Options 5 and 6 are really non-starters, as tea breaks are hardly the kind of statistic one would expect a builder's costing department to keep. The estimator therefore gets no feedback on this cost in relation to turnover.

We now discuss the estimating of each of the factors in Table 2.1.

Earnings (WR6)

Hourly rates of pay

The first thing to note is that the basic working week is of 39 hours, that is 8 hours per day on Monday to Thursday and 7 hours on Friday. The hours of work should fall within the period from 7.30 a.m. to 5.00 p.m. The 39 hours exclude a meal break of a minimum of half an hour and a maximum of one hour. Tea breaks in the morning and afternoon do not officially exist any longer, and 'the taking of tea at the place of work is not considered a breach of the present [Working] Rules' (WR6.2).

This idea of the basic week can be varied should the contract require shift work or night work, and then WR8 and WR9 are invoked.

Rates of pay are normally quoted per week and divided by the number of hours in the basic week. The reason for this is quite simple. While the basic working week for normal working is 39 hours, shift work is normally 8 hours per shift, with ten shifts in double day shift working and fifteen shifts in three-shift working. This gives 40 hours in the basic week for each operative on any shift. That the hourly rate is slightly less in this case is more than compensated by the shift allowances (see WR8.2.1 and WR8.2.2).

At the beginning of 1991, the regional rates of pay for craftsmen and labourers were as follows:

	Craftsmen		Labourers	
	Week	Hour	Week	Hour
Grade A and Scotland	£131.820	£3.380	£112.320	£2.880
London and Liverpool	£132.015	£3.385	£112.515	£2.885

Of course a builder who is only willing to pay these basic hourly rates either will not be able to attract employees, or will attract only operatives whose performance is poor in quality and quantity, or will have to have very good productivity deals or bonuses to compensate for low basic wages. Builders are forced by the shortage of skilled workers to pay in excess of these rates. The additional costs involved vary from region to region and from builder to builder, but are affected by the amount of work available, the number of builders bidding for that work, and their individual success levels. The actual rates paid in practice may be augmented by a few pence, but could well be doubled where work is plentiful and labour short. The calculations used in this text will be based on the basic rates.

Guaranteed minimum weekly earnings

So far we have discussed rates of pay as if all operatives were to be fully employed. This is not always the case. A number of factors can cause work to cease, if not for the whole building site then for parts of it or for particular operations. Such factors include excessively wet or frosty weather delaying concreting or bricklaying; high winds preventing work on roofs and high scaffolds or forcing the shutdown of tower cranes; non-delivery of essential materials; and strikes by other operatives, suppliers or subcontractors. Where these factors are not the fault of the operatives, and as long as they present themselves for work at the correct time and place, then the builder must pay at least basic rates of pay for all the hours concerned. This is known as guaranteed minimum weekly earnings. If an operative has no work for a complete week through no fault of his own, then he must receive the basic weekly rate plus the guaranteed minimum bonus (see the next section).

Obviously operatives being non-productive in this way can cause the builder considerable financial loss. It is usual when building up the cost of labour to allow for a certain amount of disruption on site due to things like weather, especially when building work is to take place in an exposed area or at times of the year when snow, rain or frost are likely. The allowance is made by a percentage addition to the all-in hourly rate and varies according to the weather pattern for the time of year, the type of work, and local conditions regarding things like deliveries.

Guaranteed minimum bonus (WR1.6)

The payment of bonus, the use of some incentive scheme by the builder or the setting up of productivity agreements, is not included in this book. It should be noted that Department of the Environment statistics on earnings and hours worked put the average bonus payments at around 30 per cent of basic wages.

One item which is included in the calculations later in the chapter is the payment of a guaranteed minimum bonus (GMB). This sum is included in the calculations since we are assuming that our fictitious contracts have no bonus or similar scheme (see WR1.6). At the beginning of 1991 the GMB payable per week for all regions was £16.380 for a craftsman and £14.040 for a labourer.

Overtime (WR7)

The special cases of shift work and night work are dealt with in WR8 and WR9.

Once an operative has worked the normal 39 hours in a week, any further working hours are designated overtime. In general terms the overtime hours worked are multiplied by a factor and the basic hourly rate is then applied. This multiplier varies according to when the overtime hours are worked, which in turn hinges on the definition of the normal working day within which the basic hours are worked.

For example, a builder may wish his operatives to work from 8.00 a.m. to 6.00 p.m. on Monday to Friday. The normal starting time is 8.00 a.m. and the lunch break will be of half an hour duration. The operatives will be working the 39 hours basic time from 8.00 a.m. to 4.30 p.m. on Monday to Thursday and 8.00 a.m. to 3.30 p.m. on Friday. Overtime will therefore have to be paid for the periods in excess of these times, i.e. for 1.5 hours on Monday to Thursday and 2.5 hours on Friday. The reader should learn to *tabulate* such information; he or she will appreciate immediately that simply writing it out can lead to errors in examinations or in practice.

Table 2.3 shows the data for this example. The rows with start, lunch and finish times are obvious. The next row is the total hours worked: subtract start time from finish time less lunch break. (This is very easy with the 24 hour clock, which is used in the examples throughout this chapter.)

The basic hours come from WR6.1 and are subtracted from the total hours worked to give the overtime hours worked. The next row is the multiplier. To find the multiplier look at WR7.5: 'Overtime shall be paid at the following rates: from Monday to Friday from agreed stopping time until 9.30 p.m., time and a half.' This means that for each hour worked, one and a half hours will be paid: a multiplier of 1.5. To obtain the overtime hours paid, take the overtime hours and apply the multiplier. To find the total hours paid at basic rate, add the basic hours and the overtime hours paid.

Productive hours are the same as total hours worked. This total is divided into the total weekly cost (to the builder) of the operative to determine the true cost of each hour when output is achieved. Non-productive hours are hours paid less productive hours. This time is called 'non-productive' because the operative is paid for it but the builder receives no work in return.

Table 2.3 might look a little pedantic and the authors would not advocate its literal use. However, something of the kind can be of great help in setting the scene, not just for overtime but for other forms of time based payment such as travelling time (see the section 'Distance to site' to follow).

WR7 contains special provision for multipliers for hours after 21.30 and for Saturday

Table 2.3   *Example of basic and overtime hours worked*

	Monday	Tuesday	Wednesday	Thursday	Friday	Total
Start	8.00	8.00	8.00	8.00	8.00	—
Lunch	0.5	0.5	0.5	0.5	0.5	—
Finish	18.00	18.00	18.00	18.00	18.00	—
Total hours worked	9.5	9.5	9.5	9.5	9.5	47.5
Basic hours	8.0	8.0	8.0	8.0	7.0	39.0
Overtime hours	1.5	1.5	1.5	1.5	2.5	8.5
Multiplier	1.5	1.5	1.5	1.5	1.5	—
Overtime hours paid	2.25	2.25	2.25	2.25	3.75	12.75
Total hours paid at basic rate	10.25	10.25	10.25	10.25	10.75	51.75
Productive hours	9.5	9.5	9.5	9.5	9.5	47.5
Non-productive hours	0.75	0.75	0.75	0.75	1.25	4.25

afternoon and Sunday. They are all quite straightforward and an accurate result can be guaranteed if a tabular method is followed.

Shiftwork and night work (WR8, WR9)

Shiftwork and night work will not be dealt with in detail in this text. Once the basic calculations involving overtime etc. are mastered, the reader should not find great difficulty in interpreting these two rules.

Travel to work (WR14, WR15)

WR14 covers travelling, and includes rules for the calculation of payment for excessive time and expense in travelling to work. A distinction is drawn between two classes of operatives: those employed in a shop or yard and those employed on a site.

For the first category (operatives working in a shop or yard), there is a limit on the amount for which the builder is liable. This limit is set at 10 miles daily for travelling time and the cost of 10 miles travel daily by public transport. We will not be concerned with this category, so upper limits will not be of interest to us.

For the second category there is no limit on time paid or on travelling expense, but alternative provisions are usually made when distances travelled become excessive (such as lodging allowance).

Travelling time

The excessive time spent travelling is based on distance in excess of 5 miles, in stages of miles or part stage. Should the distance be in excess of 25 miles, then time spent on the return portion of the journey in excess of 25 miles is paid for in 5 mile stages or part stage. The time allowed for each 5 mile stage or part stage is 15 minutes. A couple of examples will help clarify this rule, and here again the authors advocate the use of visual or tabular aids.

Example 2.1

A site is 23 miles distant. What will the travelling time be?

Scale of miles	0	5	10	15	20	25
Journey 23 miles						
Time outward only	0	0.25	0.25	0.25	0.25	

The total travelling time is 1.00 hours per day. The payment made to the operative is 1.00 hours × basic hourly rate × days.

Example 2.2

A site is 32 miles distant. What will the travelling time be?

Scale of miles	0	5	10	15	20	25	30	35
Journey 32 miles								
Time outward	0	0.25	0.25	0.25	0.25	0.25	0.25	
Return journey								
Time inward						0.25	0.25	

The total travelling time is $1.5 + 0.5 = 2$ hours per day. The payment made to the operative is 2 hours × basic hourly rate × days.

Reimbursement of fares

The excessive expense is based on the assumption that an operative can afford to pay 80 per cent of the basic hourly rate per *week* on public transport fares.

The calculation is very simple. For example, assume a craftsman has a basic rate of £3.08 per hour and a weekly fare of £12.50. The employer would be expected to reimburse the craftsman for £12.50 less 80 per cent of £3.08, i.e. £10.04.

Provision of transport

Rather than have operatives travel by public transport and risk delays etc., the employer may provide transport. In that case travelling time is still paid but obviously no expenses. The cost of this transport must be included in the cost of labour.

Lodging allowance

Should the site be very far away or in some inaccessible area, then daily travel might not be feasible. In that instance, the employer will put his operatives into lodgings and pay them in accordance with WR15. At the beginning of 1991 the rate was £14.05 per night. If the rate is found to be inadequate there is provision for agreement on an appropriate higher amount.

Where operatives are in lodgings, travel outward and inward would be paid for or transport provided at intervals depending on the distance. WR15 gives the details. Travelling time on all outward and inward journeys must be paid for.

The reader should note that certain regional variations of the Working Rules have distances for travelling time in kilometres.

Extra payments for responsibility (WR3, WR4)

Continuous extra skill or responsibility

WR3 covers extra payments for continuous extra skill or responsibility. This means that the operative receiving these extra payments is engaged and employed full time in one of 46 specific occupations. Currently there is a range of fourteen amounts from £5.46 per week to £23.01 per week. The amounts are coded A to N inclusive, and the 46 occupations have the code attached to indicate the extra payment due.

The majority of the occupations involve the handling of mechanical plant. A few others involve manual skills such as pipe laying and jointing or dry walling work. It should be noted, however, that dry walling work is not recognized as a separate skill or trade in Scotland, where this work is carried out by certain craftsmen.

Intermittent responsibility

WR4 covers extra payments for intermittent responsibility.

Two instances are defined. The first is for labourers erecting scaffolding up to 5 metres in height, who receive an extra payment over the labourer's basic rate of 11p per hour. The second is for operatives (other than craftsmen) dry cleaning stonework to remove protective material or discoloration, who receive an extra 36p per hour. (Both rates are current at the beginning of 1991.)

Holidays (WR10, WR11)

The holidays with pay scheme is designed to provide wages for all operatives during the periods when they are on annual holiday as described in the Working Rules. Three annual holiday periods are recognised – winter, Easter and summer – together with eight days of public holidays.

It should be noted, however, that the scheme does not provide pay for public holidays. The builder is bound by WR11.3 to pay wages for public holidays, and provision must be made to recoup this additional cost when calculating the cost of labour.

Holidays with pay scheme

The holidays with pay scheme is operated by a management company which is non-shareholding, non-profit-making and funded by weekly contributions from all building employers for each of their current operatives. Each individual contribution purchases a 'stamp' which is affixed to a card provided for each employee. This is done for each week the operative is in employment during the scheme's various accounting periods. The operative has his card stamped even when absent through sickness or injury up to a maximum of eight stamps. The card is the property of the operative; if there is a change of employer, the card is taken to the new employer and stamps continue to be affixed.

During the week prior to the holiday, the building employer must encash the stamps of all operatives in his employment and pay the money to the operatives on the last pay day prior to the holiday. The building employer has bought the stamps in the first instance from the management company, and must now apply to be reimbursed by the company for the payment made to his operatives.

At the beginning of 1991, holiday stamp costs to the building employer were £17.70 per week for both craftsmen and labourers. This includes a contribution of £1.55 towards retirement and death benefit (see below).

Public holidays

Recouping the cost of wages paid for public holidays involves the calculation of the possible loss incurred as a proportion of wages paid in total. This can be done simply as a percentage of basic hours worked, as follows:

Weeks per year	52
Annual holiday 21 days	4 weeks and 1 day
	48 weeks
	× 39 hours
	1872
Less 1 day of 8 hours	8
Working year	1864 hours
Public holidays: 8 days at 8 hours	64 hours

Thus public holidays represent 3.43 per cent of the working year. This would of course diminish as overtime was worked – but only in relation to the productive hours of overtime!

Retirement and death benefit schemes (WR19)

Together with the holiday with pay scheme, the management company operates death benefit, retirement benefit and accidental injury benefit schemes. The funds are managed as trusts. The management company also makes a contribution towards these funds per employee. Essentially the operatives are the beneficiaries should they retire, die in service or sustain injury in service.

At the beginning of 1991 death benefit was set at £9000, plus a further £9000 should the operative die because of injuries received at work or travelling to and from work.

Retirement benefit is paid in proportion to length of service. There is provision for the operative to buy back additional length of service and so increase the amount of benefit ultimately received. The operative pays for this buying back by having the building employer deduct the sums from his pay, which are then transmitted to the trustees of the scheme. The cost to the building employer is of course the administration of these deductions and payments.

Accidental injury benefit is payable for specified injury or injuries sustained at the place of work or while travelling to and from work. This benefit is funded from the same source as the death benefit scheme. The amount payable depends on the injury or injuries sustained. This benefit should not be confused with the sickness and injury payments made under WR16 (see later).

Employer's contributions for national insurance

The Department of Social Security publishes a number of leaflets for employers and employees which explain in varying degrees of detail the workings of the national insurance scheme. The following information has been taken from booklets NI269 and NI268, and is not intended to be more than an overview of the rules illustrated in the examples later in the chapter.

First, both employee and employer pay contributions. There is a lower limit for both and an upper limit only for employees. What constitutes a limit is the gross wage earned. This is not necessarily the gross wage used for the calculation of pay as you earn tax (PAYE: commonly known as income tax), but the difference is so slight that for practical *estimating* purposes it will be ignored in the examples. The employer is bound by law to pay both the employee and the employer contributions to the collector of taxes along with PAYE tax, but may deduct employee contributions from gross pay. These contributions for people employed by others are called class 1 contributions. Special provisions apply to people under 16, pensioners, married women and widows, and employees with more than one job.

When an employer has an occupational pension scheme, the employees are contracted out of the national insurance scheme. For example, plumbers as a whole have negotiated an occupational pension scheme with their employers. The effect is that employees and employers pay lower national insurance contributions. Of course there is the additional expense of the private occupational pension scheme, to which both employers and employees contribute.

National insurance contributions are only payable if the employee earns a certain amount or more, and as previously stated this is the gross wage. If the gross wage exceeds the lower limit it will fall into one of a number of brackets or bands. The actual rate payable varies according to which band is appropriate. Above the upper limit (again based on gross wage) there is no further increase in employee contributions but employer contributions continue to rise. The upper and lower limits are called earnings limits. Earnings limits, the brackets and the rates of insurance payable can change, normally for a tax year (6 April of one year to 5 April of the following year).

The rates of insurance payable by both employee and employer are set out in DSS booklet CF391, 'Tables – national insurance not contracted out contributions'. There are basically two ways in which the amount payable for class 1 contributions can be calculated. The first is by reference to these tables, which give weekly wages in increments of £1.00 together with the appropriate contribution of employee and employer. The match to wages paid is always to the nearest amount lower than the actual wage; for example, if the actual wage is £140.65, choose rates against £140.00. This method is frequently adopted by wages departments using manual methods of calculating wages.

The second method is to apply a percentage to the actual wage paid. This is the more accurate method, and is the one adopted for the examples later in the chapter. It is the method used in computerized payroll systems.

The rates used in the examples later in the chapter are based on the limits, rates and brackets set out in Table 2.4. In the year 6 April 1990 to 5 April 1991 the weekly lower earnings limit was £46.00 and the upper weekly limit for employees only was £350.00. There is no upper limit for employer's contributions.

Table 2.4 *Class 1 contribution rates payable by employers for employees not contracted out (tax year 6 April 1990 to 5 April 1991)*

Earnings bracket: weekly gross wage (£)	Employer's rate
46.00 to 79.99	5%
80.00 to 124.99	7%
125.00 to 174.99	9%
175.00 to 350.00	10.45%
Over 350.00	10.45%

Injury and sickness (WR16)

The Social Security and Housing Benefits Act 1982 gives every qualifying employee the right to receive statutory sick pay when unable to work because of sickness or injury. The Act lays down who will qualify, what minimum amount they will receive and for how long. Basically any operative between 16 and pensionable age qualifies if off sick for four or more consecutive days, which can include Saturdays, Sundays and holidays. The qualifying days for statutory sick pay, and a number of other things such as notification and certification of sickness or injury, have to be agreed by employer and employee. WR16 defines the qualifying days as Monday to Friday inclusive. WR16 must therefore be read in conjunction with the Act.

Sickness and injury payment

From the estimator's point of view, the most important aspect of WR16 is that it stipulates an *additional* amount of sickness and injury payment for all qualifying operatives. The qualifications under WR16 for the additional payments may vary from those required by the Act to qualify for statutory sick pay.

At the beginning of 1991 this amount was set at £55.60 per week or £11.12 per day, there being five qualifying days in the week (see WR16.3). Sickness and injury payments are only paid for a maximum of 50 working days; a minimum of four days absence is necessary before any payment is made; and no payment is made for the first three days of absence. All these rules are further bound in with the accounting period for the holidays with pay scheme and the stamping of the operative's card with a minimum number of stamps.

The bottom line for the estimator is who pays this benefit. The answer is quite simply the employer. The amount paid can vary considerably, but most standard calculations allow five sick days per year. This could be taken as a continuous period of eight days or as five separate periods of four days (three days without and one day with sick pay – twenty days sick in all). Either way the cost to the employer is the same and would be calculated as follows. From the 52 weeks in the year are deducted 21 days annual holiday and 8 days public holiday, resulting in say 46 working weeks in the year. Thus the £55.60 for five days sick pay spread over the working weeks is £1.21 per week. With 39 productive hours per week this does not produce a large sum to add to the all-in hourly rate, but it is significant enough not to be ignored. This is the amount we will allow in the examples later in the chapter.

However, this does beg the question of who pays. In the calculation above the client pays, as he always does. But what if the employer's estimator got it wrong and the employees all had six, seven or more days sick pay? In that instance the employer could lose money. He could of course insure against all sickness and injury payments, and build the insurance premiums into preliminaries or overheads or even the all-in hourly rate. He could build in five days as we have shown and insure against any excess; the premiums would be a lot less but the cost to the client at the end of the day might not be significantly less.

Statutory sick pay

Statutory sick pay must not be confused with the sickness and injury benefit above. Statutory sick pay is ultimately paid for by central government, although the employer pays the employee in the first instance and then recoups the money from central government.

Statutory sick pay can be payable for up to 28 weeks, and there is a system of linking discrete periods of sick leave to give a total of 28 weeks 'continuous' weeks in a period of three years. Among other provisions, linking depends on the periods of sick leave being no more than eight weeks apart. The employer's liability to pay statutory sick pay is limited to 28 weeks, after which the employee is transferred onto state benefit.

Statutory sick pay payments are recovered by the employer deducting all amounts paid to employees from the amounts payable for national insurance contributions. If the statutory sick pay recoverable is in excess of this amount, the excess may be deducted from the PAYE tax paid to the same collector of taxes.

Amounts payable for statutory sick pay are set out by the DSS on a weekly basis and vary according to earnings. There are three bands, namely no entitlement, standard rate or lower rate. What is paid can be topped up by the employer to some other agreed minimum, but the employer may only claim back the statutory sick pay portion paid out. This top-up effect is achieved by the provisions of WR16.

Work in difficult conditions (WR17)

The payments made under this working rule are added only to the basic hours, and not to overtime, travelling time or time lost for whatever reason. The conditions are all fully described in WR17, as is the procedure to be followed should conditions occur which have not been foreseen when the rule was framed.

The builder is expected to provide protective clothing, boots, gloves etc.

Tool and clothing allowances (WR18)

Tool and/or clothing allowances are one-off payments per week to a specific operative for the provision of his own tools and clothing. Payments should not be extended on an hourly basis to overtime, travelling time etc. Although not part of basic wages, these payments are treated as taxable earnings and are subject to deduction of PAYE and national insurance contributions.

WR18 gives lists of tools which each specific operative must provide to qualify for payment, and lists of tools which should be provided by the employer, i.e. those which it would be unreasonable to expect an operative to provide. The following crafts are involved:

Tool allowance   Woodworkers, masons, plasterers, bricklayers, slaters, wall and floor tilers.
Clothing allowance   Painters.

Special provisions are made for woodcutting machinists and apprentices.

Further provisions of WR18 cover storage of tools and clothing on site, provision of lockfast storage, insurance and liability for loss. All these items will incur an expense for the builder which must be recouped. Whereas the tool and clothing allowances are included in the cost of labour, storage and insurance are generally covered in the costing of preliminaries or treated as an overhead.

Employer's liability insurance

The Employer's Liability (Compulsory Insurance) Act 1969 makes it incumbent upon the employer to insure against injury or disease sustained by his or her operatives while at their place of work.

Premiums for this type of insurance are not expensive, and for most trades average out at around £40 per operative per annum. Glaziers, plumbers, painters and decorators and operatives engaged in road works and paving generally attract higher premiums, of the order of £80 to £100 per operative per annum. Where woodworking machinery is used, special premiums apply. Woodworking machinery in this context generally does not include hand-held power tools. The majority of policies exclude the use of explosives, tunnelling work or deep excavation, demolitions etc., for which special policies would be necessary. One can insure for literally anything – if the premiums can be afforded!

Public liability insurance (against loss or damage caused by employees) is also part of the cost of labour, and is generally offered as a package with employer's liability insurance. Although it is not a statutory requirement, many forms of building contract not only require this form of insurance but also stipulate the amount of cover required for any one event. The estimator must therefore ensure that sufficient is built in to his rates to cover these requirements.

Premiums for public liability insurance are based on a rate per operative per annum. They are generally higher than employer's liability insurance, and vary in much the same way as those for employer's liability insurance. To give £500,000 cover, the average for the majority of trades would be in the region of £50 per operative per annum. Higher rates can be expected for fencing erectors, floor layers, operatives engaged in roadworks and paving, kitchen installers, landscape gardeners and painters and decorators – somewhere in the region of £80 to £100 per operative per annum. To double the cover to £1,000,000 would not necessarily double the premium; usually an additional 20–25 per cent is all that is required. Plumbers and heating engineers are the subject of special agreement; with most companies that means higher premiums.

With both employer's liability and public liability policies there is usually a limit on the number of operatives employed. The average premiums above are for small builders with up to six operatives working on simple domestic building and repair work. Special terms apply when the work involves, for example, deep excavations or scaffolding over a particular height. There are no published rates for these types of work; the employer must obtain quotations from the insurance companies direct.

Because insurance companies generally offer both types of policy as a package, it is convenient to include both in a single percentage addition. At the beginning of 1990 the combined premium is averaging 2 per cent of the gross wage bill.

Other insurances which the employer may find it prudent to take out – covering such things as insurance of own premises, insurance of the works, and insurance of materials and plant – are not part of the cost of labour but should be included in preliminaries or as an overhead or oncost.

Construction Industry Training Board levy

This is a statutory payment to the CITB to provide training facilities for all operatives and apprentices.

Towards the end of 1990 the amounts levied were £75.00 for a craftsman and £18.00 for a labourer although these sums were under review at that time. Dividing by 46 productive weeks (as shown earlier for sickness and injury payment), the amounts we will build into the example all-in hourly rate calculations are £1.63 for a craftsman and £0.39 for a labourer.

Note that certain craftsmen such as plumbers and heating engineers attract a higher rate of levy.

Redundancy funding

All matters relating to redundancy are covered by the Employment Protection (Consolidation) Act 1978, as amended by the Employment Acts 1980, 1982 and 1988.

In general, payment of redundancy money to an operative depends on the operative being between the ages of 18 and 65 (60 for females) and having been in continuous employment with the employer for at least two years. The nature of employment in the building industry is such that few operatives are in the latter category. Labour tends to move from employer to employer as contracts are awarded and incentive schemes set up to attract operatives. Although many large contracts may run for as long as three years, it would be unusual to have the majority of operatives there for the full contract period.

The scale of payments for redundancy is based upon the weekly wage, which has a maximum of £164.00 where dismissal occurs on or after April 1988, and upon three age bands:

- 41–64 for men and 41–59 for women, for which the payment is $1\frac{1}{2}$ weeks' pay for each year of reckonable service
- 22–40, for which the payment is 1 week's pay for each year of reckonable service
- 18–21, for which payment is $\frac{1}{2}$ week's pay for each year of reckonable service.

A year is a full calendar year, and reckonable service is limited to the last 20 years of service with the employer. There are rules for the calculation of a week's pay, subject to the maximum mentioned above.

The Acts have always allowed some clawback of redundancy payments by the employer from central government funds. The amount of clawback has steadily diminished with successive amendments, and at the beginning of 1990 stands at 35 per cent of the standard redundancy payment but only for employers with ten or fewer employees. The total of redundancy payments could fall therefore upon the employer, who normally covers for this eventuality by means of a special insurance policy. Approximately 1.5 per cent of gross wages would be a reasonable allowance under current legislation.

Supervision: foremen, chargehands, gangers

The terms 'foreman' and 'chargehand' have been used indiscriminately for so long that it is difficult to know if there was ever any difference in function for these operatives. The common link is of course the supervision of operatives.

The *OED* defines foreman as 'principal workman supervising others' and chargehand as 'workman, shop assistant etc. in charge of others'. Ganger is defined as 'foreman of gang of workmen'. These definitions are more or less the same. For the purpose of this section we will consider foreman and chargehand to be synonymous, and to be a principal workman in charge of a squad of craft operatives and labourers. The ganger we will define as the principal labourer in charge of a squad of labourers. We will also adopt the terminology used in WR2.2 and refer to them as trades chargehands and gangers respectively.

No matter what we call them, these operatives must receive an additional payment as part of their wages. This additional amount is paid for every hour they work, including travel time and overtime. WR2.2 states that trades chargehands and gangers will be paid not less than £10.725 per week above the base rate for craft operatives or labourers. This works out at $27\frac{1}{2}$p per hour.

A difference we will impose is that a trades chargehand, while expected to do productive work, may have part of his time taken up with duties on the site which are non-productive. Such duties might include measuring and agreeing bonus for his squad with the builder's surveyor; ordering materials, tools or plant through the site office or through the head office; and checking materials onto site. The ganger will be expected to work productively all the time, the bonus etc. all being handled by a trades chargehand. This is not entirely satisfactory as it does not cover every situation which might arise. However, it does give a reasonable basis with which to produce sample calculations.

How much time any trades chargehand will spend on supervision will depend on the nature of the job. On large sites with many squads a trades chargehand could spend little time on non-productive work, as there would be a site agent (perhaps with assistants), various clerks ordering materials etc., and full time bonus surveyors and/or clerks. So in this case he would only agree bonus, take instruction from and report to the site agent, and keep an eye on his squad; he could be quite productive.

On small jobs, his might be the only squad. He would have to cope with everything on the site, and so would be left with little time for productive work.

Site agents are of course supervisory, but costs arising from their employment are generally covered as overheads or in the preliminaries.

So where does the cost of supervision come from for trades chargehands and gangers? For the trades chargehand it comes from two components of his wage:

- The additional supervisory payment per hour for all basic, overtime and travelling time hours.
- The payment at basic rate of all productive and non-productive hours taken up with supervision. The amount will be assessed as a percentage of hours; in Example 2.4 the trades chargehand is taken as 75 per cent productive and 25 per cent supervisory.

For the ganger it comes only from the additional payment per hour, as he is generally reckoned to be 100 per cent productive.

Tea and meal breaks (WR6)

Morning and afternoon tea breaks have already been discussed under 'Earnings'. Operatives are of course still paid during these breaks, and if 20 minutes is lost each day this amounts to a 4.26 per cent loss in production.

Lunch breaks are not paid for and so no loss is incurred.

When overtime is worked Monday to Friday, all hours in excess of normal working hours are paid at overtime rates. If there was to be a prolonged period of overtime each evening then a meal break would have to be allowed and *paid for* as overtime. The time allowed for such meal breaks has to be agreed between the building employer and operative representatives. Overtime on Saturday and Sunday might also involve meal and tea breaks. It should be noted that the cost of breaks to be recouped is increased considerably owing to the large increase in non-productive time, especially after noon on Saturday.

Bonus, incentive and productivity schemes

These are all devices to promote a higher than normal output from a workforce. For example, bonus is based on the idea that if a piece of work is completed by a group of operatives in say 20 hours rather than 30 hours, then the employer will share the cost of the 10 hours saved with the operatives involved. Therefore, to be able to pay a bonus, the employer must have used realistic outputs when calculating the bill rates. Output is defined as the number of operative hours taken to complete a unit of work.

So such schemes do not form a direct part of the cost of labour calculation for our purposes.

Excess overtime allowance

This allowance is designed to compensate for the drop in productivity consequent upon fatigue brought about by longer than usual working hours. This occurs especially when these longer hours continue for a number of consecutive days and there is also a prolonged travel period. A drop in productivity could be compensated for by adjusting the output per man in the bill rate. However, this is not a good idea as the estimator might be pricing for a number of squads doing the same work and could confuse travelling squads and local squads. These would have different all-in hourly rates anyway, so why confuse matters by assigning them different outputs? Better to price each squad realistically and leave the outputs alone.

Practical estimating

The explanatory text in this chapter had by necessity to be brief and somewhat general. The reader must appreciate that the subject of labour costs is very wide, and each topic can be studied to a greater or lesser depth. Indeed, some of the topics skimmed over here form the basis for complete careers in themselves.

However, we did say right at the beginning that for estimating purposes a few of the details need not be considered, and one can generalize for a wide range of others. This, as well as the Working Rules themselves, is best illustrated with a number of examples. We will commence with a very basic situation and proceed to more complex ideas.

Example 2.3

A squad comprising a trades chargehand (T/C), four bricklayers and four labourers is sent to a job 1 mile distant. The chargehand will work full time, i.e. supervision will be by the employer himself. The squad will work the basic 39 hours. Fares are 25p per journey.

Calculate the all-in hourly rate for craftsmen and labourers.

	Craftsman	Labourer
Basic weekly wage	£131.820	£112.320
Guaranteed minimum bonus	16.380	14.040
Tool allowance	0.860	
Gross wage	149.060	126.360
Supervision:		
T/C 39 hours at £0.275: £10.725 divided by 9 men	1.192	1.192
Ganger 39 hours at £0.275: £10.725 divided by 4 men	—	2.680
	150.252	130.232
Fares: 10 journeys at £0.35: £3.50		
Deduct 0.8 × hourly rates of £3.380 and £2.880	0.796	1.196
National health insurance (Table 2.4): 9% of gross wage	13.415	11.372
Holidays with pay	17.700	17.700
Public holidays: 3.43% of gross wage	5.113	4.334
Sick pay	1.210	1.210
Employer's liability insurance and redundancy funding:		
say 3.5% of gross wage	5.217	4.423
CITB levy	1.630	0.390
Gross cost	195.333	170.857
Tea/meal breaks: 4.26% of gross cost	8.321	7.279
Total cost per week	203.654	178.136
Divide by 39 productive hours: all-in hourly rate	£5.222	£4.567

Gross wage

The gross wage comprises only the basic wage, the GMB and the tool money allowance (see the note with WR18.1).

Supervision

The trades chargehand works or is productive all the time he is on site, the supervision being carried out by the employer. Therefore only the plussage he is paid for being a trades chargehand is chargeable to supervision. This cost is spread over all productive members of the squad, so the divisor is 1 trades chargehand + 4 craftsmen + 4 labourers = 9 men.

The appearance of a ganger in the calculation may be a surprise. It should be assumed that, where a group of labourers is to act independently, one of them at least will get a plussage for being in charge of the others. Note that this is over labourers only, so the divisor is 4, the ganger always being considered fully productive.

In the case of both trades chargehand and ganger, the additional £10.725 they are paid each week should attract NHI contributions from the employer. This would be at 9 per cent and amount to 0.965p per week, divided by 39 productive hours; it promptly disappears! It is not worth worrying about. There is some rounding up later which more than compensates for this small loss.

Fares

A very modest amount of reimbursement of fares is made and was probably not worth calculating in this case. Later examples will show how fares reimbursement can be a major expense.

National health insurance

The gross wage calculated here may not coincide with the definition of gross wage for NHI payments but is near enough for estimating purposes. The amount indicates that employer's contributions are levied at 9 per cent. Note that we are not interested in employee contributions.

Other costs

Holidays with pay are the prescribed lump sums and are always included.
   CITB levy and sickness and injury pay are included at the amounts calculated.
   Employer's liability insurance and redundancy funding are insurance premiums based on 2 per cent and 1.5 per cent of gross wage respectively, and are lumped together here for ease of calulation as 3.5 per cent.

All-in hourly rate

The total cost per week is divided by the 39 productive hours worked to obtain an all-in hourly rate.

Example 2.4

A squad comprising a trades chargehand (T/C), four craftsmen (bricklayers) and four labourers is sent to a job 21 miles distant. The T/C will work 75 per cent of the time, i.e. supervision will take 25 per cent of his time. The squad will work overtime 2 hours/day Monday to Thursday. Fares are £2.85 return. Include an allowance of 7.5 per cent for guaranteed minimum weekly wage.
   Calculate the all-in hourly rate for craftsmen and labourers.

		Craftsman	*Labourer*
Basic weekly wage		£131.820	£112.320
Overtime:			
Craftsman: factor 1.5 × 4 days × 2 hours at £3.380		40.560	–
Labourer: factor 1.5 × 4 days × 2 hours at £2.880		–	34.560
Travel time: 21 miles, divide by 5, less 1, round up: 4			
4 × 0.25 hours × 5 days: 5 hours at £3.380 and £2.880		16.900	14.400
Guaranteed minimum bonus		16.380	14.040
Tool allowance		0.860	–
Gross wage		206.520	175.320

Supervision: trades chargehand	
Hours: basic 39, overtime 8 + 4, travel 5: 56 at £0.275	£15.400
Craftsman's gross wage	206.520
T/C gross wage	221.920
Fares: 5 at £2.85: £14.25 less 0.8 × hourly rate £3.655	11.326
NHI: 10.45% of gross wage	23.191
Holidays with pay	17.700

Public holidays: 3.43% of gross wage	7.612	
Sick pay	1.210	
ELI and redundancy fund: 3.5% of gross wage	7.767	
CITB levy	1.630	
T/C gross cost	292.356	
Tea/meal breaks: 4.26% of gross cost	12.454	
Guaranteed week allowance: 7.5% of gross wage	16.644	
Total T/C cost	321.454	
Less hourly plussage	15.400	
	306.054	
Take 25% as supervisory	76.514	
Add hourly plussage	15.400	
	91.914	
Divide by 8.75 men supervised	10.504	10.504

Supervision: *ganger*

Hours 56 at £0.275: £15.40 divided by 4 men	–	3.850
Fares: 5 journeys at £2.85: £14.25		
Deduct 0.8 × hourly rates of £3.380 and £2.880	11.546	11.946
National health insurance (Table 2.4): 10.45% of		
gross wage	21.581	18.321
Holidays with pay	17.700	17.700
Public holidays: 3.43% of gross wage	7.084	6.013
Sick pay	1.210	1.210
Employer's liability insurance and redundancy funding:		
say 3.5% of gross wage	7.228	6.136
CITB levy	1.630	0.390
Gross cost	285.003	251.390
Tea/meal breaks: 4.26% of gross cost	12.141	10.709
Guaranteed week allowance: 7.5% of gross wages	15.489	13.149
Total cost per week	312.633	275.248
Divide by productive hours of 39 basic + 8 overtime = 47:		
all-in hourly rate	£6.652	£5.856

Hours paid

The total number of hours paid for each man in this example is made up of 39 basic hours, 8 overtime hours, 4 additional hours obtained by taking 0.5 × overtime hours, and 5 hours of travelling time: this gives 56 hours total. Of these hours, 47 are productive and 9 are non-productive.

Travelling time

An alternative method of calculating the travelling time of one hour per day is as follows.

Scale of miles	0	5	10	15	20	25
Journey 21 miles	———————————→					
Time outward only	0	0.25	0.25	0.25	0.25	

Supervision

The calculation for supervision for the trades chargehand seems very complex. However, all that is done is the following:

1 We calculate the T/C's gross wage by taking the craftsman's gross and adding the hours worked at the T/C's plussage.
2 We then calculate everything else as would be done for craftsmen and labourers, only using the T/C's hourly rate and the T/C's gross wage.
3 When we have the total cost for the T/C, we separate the additional cost per hour (which is *all* a supervisory cost) from the remainder (which is only 25 per cent supervisory cost).
4 When we arrive at the latter figure we determine what constitutes 25 per cent and add back the hourly plussage.
5 Finally we divide by the number of men supervised. This is made up of the 4 craftsmen, the 4 labourers and the 0.75 of the T/C which is productive, giving 8.75.

National health insurance

Note that there are two different calculations for NHI contributions since the gross wages of craftsmen and labourers fall within different brackets.

Guaranteed week allowance

First let us clarify a point. This is not a 7.5 per cent loss of production, costs or anything else. If it was a loss it would not be mathematically correct to *add* it to our rate; we would expect to add 8.11 per cent.

There is a difference of opinion about the amount to which we should apply the 7.5 per cent. Some schools of thought add it to the basic wage costs only, since that is all that is paid for the guaranteed weekly wage. Others add it to gross wage, as that is the cost to be recouped. It does, however, depend on how the percentage has been arrived at in the first place. If the cost of the guaranteed weekly wage is recorded for each trading year or even individual contract, then it will be expressed as a percentage overhead of some fixed sum. The latter can be anything within reason, such as the total basic wage bill, the gross wage bill or the gross wage costs; it makes no difference as long as the estimator knows which.

As a matter of interest, if we take the guaranteed week allowance as 7.5 per cent gross cost in the example, the difference on the craftsman rate would be £21.375 − £15.489 = £5.886; divided by 47 hours, this gives £0.125. Twelve and a half pence an hour is quite a difference, so the estimator has to know what base is used by the cost accountant when he is given these percentages.

Example 2.5

A squad comprising a trades chargehand (T/C), three craftsmen (bricklayers) and two labourers is sent to a job 38 miles distant. The T/C will work 75 per cent of the time, i.e. supervision will take 25 per cent of his time. One labourer operates a 10/7 concrete mixer, the other a mechanical elevator. The squad will work overtime 3 hours/day Monday to Thursday. The squad will travel out to site on Monday, return on Friday evening, and stay in lodgings Monday to Thursday nights inclusive. Fares are £6.00 return. Include an allowance of 7.5 per cent for guaranteed minimum weekly wage and 5 per cent excess overtime allowance.

Calculate the all-in hourly rate for craftsmen and labourers.

	Craftsman	Labourer
Basic weekly wage	£131.820	£112.320
Overtime:		
Craftsman: factor 1.5×4 days $\times 3$ hours at £3.380	60.840	
Labourer: factor 1.5×4 days $\times 3$ hours at £2.880		51.840
Travel time:		
Out: 38 miles, divide by 5, less 1, round up: $7 \times 0.25 = 1.75$		
In: 13 miles, divide by 5, round up: $3 \times 0.25 = 0.75$		
Total 2.5 hours times £3.380 and £2.880	8.450	7.200
Guaranteed minimum bonus	16.380	14.040
Tool allowance	0.860	
Labourers' plussage for 59.5 hours at £0.140		8.330
Gross wage	218.350	193.730

Supervision: *trades chargehand*
Hours: basic 39, overtime $12 + 6$, travel 2.5:

59.5 at £0.275	£16.363
Craftsman's gross wage	218.350
T/C's gross wage	234.713
Fares: 1 at £6.00, less $0.8 \times$ hourly rate £3.655	3.076
Lodging allowance	56.200
NHI: 10.45% of gross wage	24.528
Holidays with pay	17.700
Public holidays: 3.43% of gross wage	8.051
Sick pay	1.210
ELI and redundancy fund: 3.5% of gross wage	8.215
CITB levy	1.630
T/C gross cost	355.323
Tea/meal breaks: 4.26% of gross cost	15.137
Guaranteed week allowance: 7.5% of gross wage	17.603
Excess overtime allowance: 5% of gross cost	17.766
Total T/C cost	405.829
Less hourly plussage	16.363
	389.466
Take 25% as supervisory	97.367
Add hourly plussage	16.363
	113.730

	Craftsman	Labourer
Divide by 5.75 men supervised	19.779	19.779
Fares: 1 journey at £6.00		
Deduct 0.8 × hourly rates of £3.380 and £2.880	3.296	3.696
Lodging allowance	56.200	56.200
National health insurance (Table 2.4): 10.45% of gross wage	22.818	20.245
Holidays with pay	17.700	17.700
Public holidays: 3.43% of gross wage	7.489	6.645
Sick pay	1.210	1.210
Employer's liability insurance and redundancy funding:		
say 3.5% of gross wage	7.642	6.781
CITB levy	1.630	0.390
Gross cost	356.114	326.276
Tea/meal breaks: 4.26% of gross cost	15.170	13.899
Guaranteed week allowance: 7.5% of gross wages	16.376	14.530
Excess overtime allowance: 5% of gross cost	17.806	16.314
Total cost per week	405.466	371.019
Divide by productive hours of 39 basic + 12 overtime = 51:		
all-in hourly rate	£7.950	£7.275

Travelling time

The calculation for the travelling time for 38 miles is best illustrated graphically. The essential point is that any miles over 25 attract hours for a return journey.

Scale of miles	0	5	10	15	20	25	30	35	40
Journey 38 miles	→								
Time outward	0	0.25	0.25	0.25	0.25	0.25	0.25	0.25	
Return journey	←								
Time inward							0.25	0.25	0.25

The total travelling time is 1.75 + 0.75 = 2.5 hours. In this example the journey is weekly rather than daily.

Plussages

The labourer's plussage is 14p per hour; both skills are listed as an A rate in WR3.

No ganger plussage is taken in this example, as the labourers are part of a bricklaying squad and are therefore under the direct control of the trades chargehand.

3

Mechanical plant

Introduction

Mechanical plant operated by the builder falls into one of two categories:

● Small plant and tools
● Large mechanical plant and scaffolding.

While either category can be owned by the builder or hired from a plant hire company, small plant and tools such as hand-held power drills and saws are normally purchased by the builder. The builder will cover the costs of such tools by a percentage addition to the contract. The method for the incorporation of this cost is described in Chapter 4, 'preliminaries'.

The estimator, in pricing an item in the bill of quantities which includes a plant element, will use either the hire charges quoted by a plant hire company or a rate which has previously been calculated for company owned plant. The plant rate will normally be quoted as a rate per hour either including or excluding the operator. Simplistically, if it is assumed that a hydraulic excavator with a 1 m^3 bucket can carry out 30 operations per hour and costs £45 per hour, then the excavation costs with this machine will be £1.50 per m^3. There are a number of factors associated with this example, such as how the £45 per hour was derived, the number of operations per hour, the disposal of the excavated material and the balancing of the excavator with the disposal plant. These factors are the subject of this chapter.

Purchase or hire

Hired plant

Plant can be obtained in a variety of ways, the particular method being a reflection of the builder's use of the plant. Generally a builder will own plant which is used daily, whereas plant which is required for a particular operation will be hired on a short term basis from a plant hire company. This latter category will tend to include specialist plant such as mobile cranes. The hire charge terms of a plant hire company will normally include all costs except fuel, insurance and transporting the plant to site. Whether or not an operator is hired with the plant will depend upon the type of plant; obviously for very specialist plant the builder will hire the plant together with an operator.

Period hire

In situations where a builder requires plant or vehicles for a longer period but does not wish to own them, there are other arrangements available. For example, a builder is to commence a two year project which requires one van and one truck as site transport. The builder cannot foresee a use for these vehicles after the end of the contract. One option is to hire as above but an alternative is one of the following.

Contract hire

This is defined as a monthly contract for the hire of the plant or vehicle which covers everything except the fuel, insurance and operator. Contract hire is carried out by specialist contract hire companies who agree to hire for a period that is not necessarily the life of the plant or vehicle.

Leasing

This is similar to contract hire but excludes maintenance of the plant or vehicle. The builder will pay a monthly charge for the use and will pay for maintenance. The builder will be able to lease for a period which is less than the life of the plant or vehicle. The Inland Revenue imposes restrictions on the minimum and maximum periods of the lease. Very often under leasing the vehicle is sold at auction at the end of the lease period. In this case there is a reserve price in the lease agreement which is the assumed value that the vehicle will realize at auction. If the vehicle realizes less then the builder will have to make a cash payment to the leasing company. If the vehicle realizes more then the excess is passed to the builder.

Builder's own plant

There are a number of ways in which the builder can purchase plant.

Outright purchase

This is the most straightforward method. It involves the builder making a purchase out of cash reserves. The calculation of the real cost of this action is not so straightforward, however, since by investing in a piece of plant the builder has to forgo the opportunity of leaving his money on deposit in a bank or even financing another construction project. It can be argued that the cost of outright purchase is the interest which is lost by not putting the money on deposit in the bank. There is also another dimension: if the money is put on deposit in a bank, then after some years the original sum can still be withdrawn. If however the money is invested in a piece of plant then, as all car owners appreciate, eventually the builder will possess a worthless heap of scrap. The cost of outright purchase is therefore the interest lost plus the sum of money which would be required to be invested each month to amount to the value of the purchase price over the life of the plant. This latter investment is termed a sinking fund.

Bank or finance company loan

A bank or finance company loan will be made to the builder for the purpose of purchasing plant. The loan may be for the full or part cost of the plant. The lender may require that the builder puts in part of the cost from his own reserves, and will certainly demand that the loan is secured against the builder's capital assets as a whole.

Hire purchase

Under a hire purchase agreement the builder undertakes to repay a loan made by a hire purchase company over a period. The builder will be required to place a deposit, the amount of which may be governed by legislation. The hire purchase company secures the loan against the plant item being purchased, and in the event of a default has no direct recourse to the builder's assets as a whole.

Lease purchase

Lease purchase is similar to leasing except that the plant is owned by the builder and appears on the balance sheet. This method of financing takes the total value of the plant down to zero by virtue of the monthly payments, and the title to the vehicle is then transferred on payment of a nominal sum. Also the contract may be terminated at any time during the period of the agreement by payment of a termination charge, which is calculated on a sliding scale over the period of the agreement. There is normally in force a legal requirement that the initial payment be a number of months of lease in advance. Under this method the value of the plant is normally written down, i.e. depreciated, over the period of the lease such that the plant has a nominal value on the balance sheet at the end of the lease.

The plant rate

Mechanical plant for construction work may be hired or purchased in one of the ways described above. Either the plant hire company or the contractor must calculate the cost of plant, which is normally on a rate per hour basis. This section will describe the factors to be taken into account. There follow examples illustrating the procedure.

The plant rate is built up from the following factors:

- Initial cost and finance
- Depreciation
- Interest on capital borrowed
- Life of plant
- Hours worked per annum
- Repairs and renewals
- Insurance and licences
- Fuel, oil and grease
- Inflation.

In addition the estimator must also consider the cost of

- Transporting the plant to site, which will obviously depend upon the relative positions of the plant and the site
- The operator, which will vary with such factors as travelling time, bonuses etc. (see Chapter 2).

Initial cost and finance

There are two ways to consider the purchase of plant. The first is that the purchaser has no money and therefore has to borrow to purchase using one of the methods described above. The second is that the purchaser has sufficient money to purchase the plant out of his own reserves.

Borrowing

In the first instance the purchaser has to borrow money from a bank or finance company in order to purchase the particular piece of plant. The cost is therefore the cost of the purchase plus the cost of the loan.

Assume that an item of plant has been purchased for £100,000 and that it has a life of seven years. Assume also that the purchaser is able to raise a mortgage, either on that plant for its full value or on other assets such as the purchaser's own office building. The cost to the purchaser would be based upon the mortgage equation, which is

$$R = P[(1+i)^n i]/[(1+i)^n - 1]$$

where P is the principal borrowed, in this case £100,000; i is the interest rate as a decimal; n is the borrowing period in years, in this case seven years; and R is the annual repayment.

If it is assumed that a bank is willing to lend at 13.5 per cent, then the rate of interest as a decimal is 0.135. Thus the annual repayment is given by

$$R = £100,000[(1.135)^7 0.135]/[(1.135)^7 - 1] = £22,964$$

Therefore the cost over seven years is £160,748, and the interest paid is £60,748.

Purchasing from reserves

In the second case it is assumed that the purchaser has the money and therefore the cost is a simple £100,000. However, this ignores the fact that the purchaser could invest the money in the bank rather than in a piece of plant. The purchaser therefore suffers the loss of the interest on the investment. Moreover, had the purchaser invested the money in the bank then at the end of the seven years the full £100,000 could have been withdrawn, whereas investing in plant results in the possession of a rusty heap of scrap In this case the cost over seven years is the loss of interest plus the contribution to the sinking fund, which accumulates through annual investment to £100,000 over seven years.

If it is assumed that the rate of interest for investments is 7.5 per cent, then the annual interest lost is £100,000 × 0.075 = £7500.

The annual sinking fund equation is

$$R = Ai/[(1+i)^n - 1]$$

where A is the target accumulated amount, in this case £100,000; i is the interest rate as a decimal, i.e. 0.075; n is the investment period, in this case seven years; and R is the investment expressed as an annual payment. Thus the annual contribution to the sinking fund is

$$R = £100,000 \times 0.075/[(1.075)^7 - 1] = £11,380$$

The total cost per annum is therefore £7500 + £11,380 = £18,880. Hence the cost over seven years is £132,160.

The rate assumed here is based upon the percentage rate given on deposit in a bank. It may be argued, however, that the percentage rate which should be used is the actual rate of return of investing in the builder's company, since by buying plant the builder is losing the opportunity of investing in his own company. The company rate of return for each £1 invested may well be 15 or even 20 per cent, and therefore this is the rate which should be used in the above calculation. It will be noted that the taxation implications for the sinking fund are ignored altogether here, whereas in practice they will be taken into account.

It is up to individual preference as to which method is used to calculate cost and finance. However, the former, assuming no capital, has a lot to commend it in terms of its ease of calculation and its logic.

Depreciation

There are two common ways of calculating depreciation for the purposes of the plant hire rate, namely the straight line method and the written down value method. Of the two the straight line method is to be preferred for the purposes of calculating a plant hire rate. It is worth noting here that while all aspects of the plant hire rate may currently be set against corporation tax, the purchase of plant may not. Since the recent abolition of the capital equipment allowance, accountants will set the initial expenditure made on the plant as a company expenditure from taxed income. The value of the plant is included in the assets of the company. Depreciation is taken as an expenditure in each year that the plant is depreciating.

Straight line method

This method simply involves the estimation of two factors:

● The economic life of the particular plant
● The residual or scrap value of the plant at the end of its economic life.

If the plant costs £100,000 and has a scrap value of £10,000 after seven years, then the amount of depreciation per year is calculated as follows:

Cost of plant	£100,000
Value at end of economic life	− 10,000
	90,000
Years	÷ 7
Depreciation per year	£12,857

Written down value method

Under this method an unchanged percentage rate (commonly 25 per cent) is deducted as depreciation from the previous year's value. This method has the advantage that it more accurately reflects the resale value of the plant at any time, because plant loses the majority of its value in the first year.

Using the same figures as above:

Capital cost of plant	£100,000
Less 25% depreciation in first year	25,000
Remaining value of plant	75,000
Less 25% depreciation in second year	18,750
Remaining value of plant	56,250
Less 25% depreciation in third year	14,063
Remaining value of plant	42,187
Less 25% depreciation in fourth year	10,547

Remaining value of plant	31,640
Less 25% depreciation in fifth year	7,910
Remaining value of plant	23,730
Less 25% depreciation in sixth year	5,933
Remaining value of plant	17,797
Less 25% depreciation in seventh year	4,449
Remaining value of plant at end of seventh year	£13,348

This method, although used in accounting, is not generally suitable for deriving a plant hire rate since it presumes a new calculation at the end of each year.

Interest on capital borrowed

The assessment of the total amount of interest has been covered at length above under the heading 'Initial cost and finance'. To determine the average yearly value of interest on a loan, the total interest paid is divided by the number of years. In the above example this would be £60,748/7 = £8678.

Life of plant

The anticipated life of plant is generally established by experience. Two factors come into play. The first is the period for which the plant will physically last before the amount of the repair bills become too high compared with the value of the plant. The second is the expected period before the plant is made obsolete by advancing technology. For the purpose of determining a plant hire rate, an estimate on the cautious side is normally the basis for calculation.

Statements from the technical salesmen of plant manufacturers are often a good guide. Such statement might be: 'The plant is guaranteed for one year and all hydraulic seals for two years; most of our operators seem to get about seven years; and there are some machines still operating which are ten years old.' This could be taken to mean that there will be little need of repairs in the first two years but following that repairs will increase, until after five to seven years the plant will require to be replaced: it is unusual for it to last ten years.

Hours worked per annum

Plant is never utilized for every working hour of every working day of the year. Certain types of plant – such as concrete mixers, wheeled tractor mounted front shovels, backhoe excavators of the JCB3 type, vans and trucks – are likely to be in use for the majority of the year. On the other hand, specialist types of plant such as motorized scrapers and mobile concrete pumps will be restricted in their use by the weather or particular demand.

The number of working hours per annum for a site is approximately 1900. A general purpose excavator may be on hire for 1500 hours per annum, whereas a mobile concrete pump may be on hire for only 800 hours per annum. The exact figures will be drawn from the plant hirer's or contractor's own records.

During the on-hire period other factors will reduce the working efficiency of the plant. For example, for an excavator the following time might be lost (as a percentage of available time):

Weather	10
Manoeuvring	10
Breakdowns	5
Operator efficiency	5
General waiting	10
	40%

Such factors are taken into account when calculating the effective output of excavating plant per hour.

Repairs and renewals

All mechanical plant suffers from breakdown due to the failure of components. This has two cost effects, the first being the cost of the repair and the second being time lost. The time lost is taken into account in the working output of the plant as described above. Breakdown may be prevented by renewing worn parts during normal periodic maintenance.

The cost of repairs and renewals will vary for different types of plant, but a general figure would be 10 per cent of the purchase price per annum.

Insurance and licences

It is normal for builders today to have a contractor's all risks (CAR) insurance policy. This will cover, as the name implies, all risks associated with the construction of the building. The premium for this policy is charged as a percentage of turnover (see Chapter 4, 'Preliminaries'). The implications for plant are as follows:

Employer's liability   This cover is required under an Act of Parliament to cover all employers for their liability in respect of death or injury to workpeople. This cover is extended by CAR policies to cover operatives hired with plant.

Public liability   This insurance will cover the builder against claims by third parties. If, for example, through the negligence of the operative an excavator knocks down a boundary wall and damages a car parked on the other side, then the insurance will pay for claims by the owner of the car.

Works insurance   This insurance is primarily for a new building, but can be extended to cover plant for fire, theft and damage. Damage may be accidental or through vandalism. The insurance may also be extended to cover the cost of recovery of incapacitated plant, for example an excavator bogged down in bad ground.

The above insurances are specific to site based plant which is not licensed to use the public road. For road vehicles such as vans, trucks and mobile cranes it is necessary to have a vehicle-specific insurance under the Road Traffic Act and also a road fund licence. Since these costs are vehicle specific they are included with the plant hire rate, unlike site based plant where these costs are excluded.

Fuel, oil and grease

Fuel consumption is generally given in terms of litres per hour. Diesel is available as derv for use on public roads, and as gas oil for use on site. The latter is less expensive owing to tax concessions; it contains a red dye and should not be used in road vehicles.

The best source of fuel consumption figures is the site records. However, a useful rule of thumb is that a petrol engine will consume approximately 0.275 litres of fuel per brake horsepower (bhp) hour, and a diesel engine will consume approximately 0.2 litres of fuel per bhp hour.

From the catalogue of a common hydraulic tracked excavator normally used with backhoe equipment, the maximum power output is given as 130 bhp. When used to excavate a trench, this machine will be working for 5 seconds at full power during a cycle of 20 seconds. For the remaining 15 seconds it is likely to be operating at about half this power. The diesel fuel consumption is therefore calculated as follows:

$$0.25 \times 130 \times 0.2 = 6.5$$
$$0.75 \times 65 \times 0.2 = \underline{9.75}$$
$$16.25 \text{ litres per hour}$$

Similar information from the catalogue of a 100 litre (5/3.5) concrete mixer states that the power output is 2.5 bhp. Here the engine will work constantly under approximately the same load, and therefore diesel fuel consumption will be $2.5 \times 0.2 = 0.5$ litres per hour. This correlates well with the old approximation of 1 gallon per day.

The cost of lubricating oil and grease can be taken as 10 per cent of the fuel cost.

Tracks and tyres

The cost of maintenance to tracks on plant working on normal ground can be considered to be included with the repairs and renewals figure as calculated above. An exception to this would be where tracked plant is working on rock, for instance in a quarry, where track failure is more common. In this instance repairs and renewals should be 12 per cent of the capital cost per annum.

Tyre life is determined by the tyre size, the loading, the tyre pressure, the type of ground and the skill of the operator. Tyres can represent 15 per cent of the capital cost of a wheeled loader, and for this reason various techniques are available for prolonging tyre life. Filling the tyre with a flexible foam means that tyre 'pressure' is always correct and production losses due to punctures are removed. Fitting chains to tyres will prolong life, particularly when working on rock. Tyre life for normal and harsh operating conditions is as follows. The figure for harsh conditions should only be used when it is envisaged that the plant will be working continuously in a harsh environment.

Loaders: normal 2500 hours; harsh 1500 hours
Trucks: normal 3000 hours; harsh 2000 hours.

Inflation

A plant rate is calculated at the time that plant is purchased. If no account were to be taken of inflation then during the last year of the life of the plant the rate would be uncompetitively low. Rates are therefore reviewed on an annual basis and increases for inflation are added.

Example 3.1

Calculate the plant rate per hour for a 100 litre (5/3.5) concrete mixer based on the following information:

Purchase price	£1500
Interest rate	13.5 per cent
Engine power	2.5 bhp
Gas oil	£0.16 per litre
Life of mixer	10 years
Scrap value at end of life	nil
Hours worked per annum	1600
Working efficiency	90 per cent

Depreciation: £1500/10	£150.00

Interest on capital borrowed using the mortgage formula:
$$R = £1500[(1.135)^{10}0.135]/[(1.135)^{10} - 1] = £282 \text{ per annum}$$
Total repayment $= £282 \times 10 = £2820$
Interest $= £2820 - 1500 = £1320$

Interest per annum: £1320/10	132.00
Repairs and renewals: 10% of £1500	150.00
Insurance: none	nil
	432.00
Divide by hours per annum: £432/1600	0.27
Fuel: 2.5 bhp $\times$ 0.2 = 0.5 litres/hour $\times$ £0.16/litre	0.08
Oil and grease: 10% of fuel	0.01
	0.36
Efficiency rating 90%: £0.36/0.9	£0.40

The rate for a 100 litre concrete mixer is £0.40 per hour.

Example 3.2

Calculate the plant rate per hour for a 114 bhp wheeled backhoe excavator based on the following information:

Purchase price	£85,000
Interest rate	13.5 per cent
Engine power	114 bhp
Gas oil	£0.16 per litre
Life of excavator	7 years
Scrap value at end of life	£1000
Hours worked per annum	1400
Working efficiency	60 per cent
Tyre cost	£12,000 per set of 8
Tyre life	2500 hours

Depreciation is cost less tyre and scrap values:
 £85,000 – 1000 – 12,000 = £72,000: divide by 7   £10,285.71
Interest on capital borrowed using the mortgage formula:
 $R = £85,000[(1.135)^7 0.135]/[(1.135)^7 - 1] = £19,520$ per annum
Total repayment = £19,520 × 7 = £136,640
 Interest = £136,640 – 85,000 = £51,640
 Interest per annum: £51,640/7   7,377.00
Repairs and renewals: 10% of £85,000 – 12,000   7,300.00
Insurance: none

                    24,962.71

Divide by hours per annum: £24,962.71/1400   17.83
Tyre depreciation: £12,000/2500   4.80
Fuel, based on 25% at full power:
 114 bhp × 0.2 × 0.25 = 5.7 litres/hour
 57 bhp × 0.2 × 0.75 = 8.6 litres/hour
 Total 14.3 litres/hour × £0.16/litre   2.29
Oil and grease 10% of fuel   0.23

                    25.15

Efficiency rating 60%: £25.15/0.6   £41.92

The rate for a 114 bhp wheeled backhoe excavator to be used in calculating an excavation rate is £41.92 per hour.

Example 3.3

Calculate the plant rate per hour for an 80 bhp 5 m^3 truck based on the following information:

Purchase price	£35,000
Interest rate	13.5 per cent
Engine power	80 bhp
Derv	£0.28 per litre
Life of truck	5 years
Scrap value at end of life	£1000
Hours worked per annum	1800
Insurance and road fund licence	£1200 per annum
Working efficiency	60 per cent
Tyre cost	£3,000 per set of 6
Tyre life	3000 hours

Depreciation is cost less tyre and scrap values:
 £35,000 – 1000 – 3000 = £31,000: divide by 5   £6,200.00
Interest on capital borrowed using the mortgage formula:
 $R = £35,000[(1.135)^5 0.135]/[(1.135)^5 - 1] = £10,073$ per annum
 Total repayment = £10,073 × 5 = £50,365
 Interest = £50,365 – 35,000 = £15,365
 Interest per annum: £15,365/5   3,073.00
Repairs and renewals: 10% of £35,000 – 3000   3,200.00
Insurance and road fund licence   1,200.00

                    13,673.00

Divide by hours per annum: £13,673/1800	7.60
Tyre depreciation: £3000/3000	1.00

Fuel, based on 75% at full power:
   80 bhp × 0.2 × 0.75 = 12 litres/hour
   40 bhp × 0.2 × 0.25 = 2 litres/hour

Total 14 litres/hour × £0.28/litre	3.92
Oil and grease: 10% of fuel	0.39
	12.91
Efficiency rating 60%: £12.91/0.6	£21.52

The rate for an 80 bhp 5 m³ truck is £21.52 per hour.

Example 3.4

Using the plant described in Examples 3.2 and 3.3 and the information given below, calculate the rate per m³ for machine excavation of a trench 600 mm wide and the disposal of surplus excavated soil at a tip 2 miles from the site.

Wheeled backhoe excavator:

600 mm wide, 400 litre (0.4 m³) bucket, 1 cycle per 30 seconds	£41.92/hour
Operative	£5.00/hour

Truck 5 m³:

average speed to tip 12 mph, tipping time 2 minutes	£21.52/hour
Driver	£4.50/hour

Normal ground: bulking rate 1.1

Time taken to excavate 5 m³ of bulked earth:
  5 m³ truck capacity, divided by 0.4 m³ bucket capacity: 12.5 cycles
13 cycles of excavator to fill one truck, at 30 seconds per cycle: 6.5 minutes to fill one truck

Truck cycle time (minutes):

Loading	6.5
Travelling	20.0 (4 miles at 12 mph)
Tipping	2.0
Cycle time	28.5

Number of trucks required: 28.5/6.5 = 4.38 trucks

A choice is now available between having 5 trucks with the excavator working to capacity, or 4 trucks with the excavator standing idle for a short period. Each option is now costed.

For 5 trucks the cost per hour is as follows:

Excavator and operative	£46.92
5 trucks and drivers: 5 × £26.02	130.10
Total cost	£177.02

Output per hour (excavator): (3600/30 sec) × 0.4 = 48 m³ of bulked earth
Therefore cost is £177.02/48 = £3.69 per m³

For 4 trucks the cost per hour is as follows:

Excavator and operative	£46.92
4 trucks and drivers: 4 × £26.02	104.08
Total cost	£151.00

Output per hour (trucks): (60/28.5 min) × 5 m^3 × 4 trucks = 42 m^3 of bulked earth
Therefore cost is £151.00/42 = £3.60 per m^3.

Therefore the operation with 4 trucks is more economic.

Thus the rate for excavating a trench 600 mm wide and disposing of excavated material is £3.60 × 1.1 (bulking factor) = £3.96 + 20% overheads and profit = £4.75 per m^3.

4

Preliminaries

SMM7 Section A

Introduction

The preliminaries section of the bill of quantities is the estimator's introduction to the contract. It informs the estimator of the location, size and complexity of the project, and gives details of the conditions of contract under which the project is to be completed. The preliminaries section needs care in pricing, since it is here that the contractor is to cover the cost of operating the site under specified conditions and in accordance with the contractor's plan for the progress of the work and storage and movement of material.

For the purposes of illustration of the pricing of the preliminaries section, all examples will refer to the construction over one year of a 48 m × 56 m factory unit on an existing industrial estate.

The clause numbers following the headings are those of the *Standard Method of Measurement of Building Works*, Section A.

Project particulars (A10)

The project particulars give the name, nature and address of the site along with the names and addresses of the client and the various members of the design team.

Drawings (A11)

This section includes a list of drawings which were used in the preparation of the bills of quantities and will therefore become contract drawings. It is not normal for all the drawings to be sent with the bills of quantities to tendering contractors, and details will be given of the arrangements which the contractor may make for the inspection of drawings.

The site (A12)

Details of the site are given. For example:

> The site is situated in Southern Industrial Park, to the west of the existing superstore unit. Access to the site will be by surfaced roads from Albert Parkway and Bishop Road. The contractor will be confined to the area of the site indicated in red on drawing AA/101/89. The contractor is advised to visit the site to ascertain the nature of the site and all local conditions.

Description of the works (A13)

The works are briefly described to give the estimator a feel for the size and scope of the works and also to allow the estimator to carry out rough calculations for scaffolding etc. For example:

> The works comprise the construction of a factory unit, grid size on plan 48 m × 56 m, with a height to the underside of the roof beam at eaves of 5.72 m.
>
> Construction is steel column and beam portal frame, composite panel cladding and aluminium standing seam roofing system. The factory has a two-story office within the main structural envelope. Mechanical and electrical services installation is included in the contract, as are external works comprising roads, car parks, service yards, footpaths and landscaping.

The estimator is now in possession of all necessary basic information. From this point onwards the preliminaries become more specific and can be priced.

The contract (A20)

The list of contract conditions gives the estimator the opportunity to price the express requirements of the contract, and to estimate the cost of the risks which are to be borne by the contractor.

This section will assume the use of the Standard Form of Building Contract, Private With Quantities, 1980 edition with amendments 1,2,4,5,6,7 and 8, issued by the Joint Contracts Tribunal (abbreviated JCT80). Also, for the purposes of this section all clauses will be listed whether or not there is a pricing implication. The JCT80 clause numbers follow each heading.

Interpretation, definitions etc. (1)

Contractor's obligations (2)

Contract sum: additions or deductions (3)

Architect's or contract administrator's instructions (4)

Contract documents: provision of a master programme (5)

Although this clause requires the contractor to provide a copy of the master programme, it is probable that this would have been produced by the production planning team within head office and would therefore be costed as a head office overhead. It is thus unlikely that this item would be priced here. Head office overheads are discussed later.

Statutory obligations notices, fees and charges (6)

Where fees are payable arising out of the works described in the bills of quantities then these are to be included within the contract sum. Rates on site hutting may be payable but in this example they are included in the valuation of site hutting.

Levels and setting out of the works (7)

Materials, goods and workmanship to conform to description, testing and inspection (8)

This is covered more fully under clause A33 (see later).

Royalties and patent rights (9)

If the contractor is required to use a patented process and this is fully described in the bills of quantities, then the contractor should allow here the cost of any licence fee payable.

Person in charge (10)

Traditionally estimators used this item to price the total cost of the site administration. However, bills of quantities prepared in accordance with SMM7 will contain the item 'A40 Management and staff' under which site administration is priced (see later).

Access for architect or contract administrator (11)

Clerk of works (12)

Facilities for the clerk of works are priced under item A41 (see later).

Variations and provisional sums (13)

Contract sum (14)

Value added tax (15)

The conditions relating to VAT are included in a supplement to JCT80 and also in amendment 8. The contract sum and valuations are net of VAT. The contractor can recover VAT on materials purchased and must charge on certain types of building work. At the time of writing only new buildings are zero rated, all other building work being subject to VAT. This could easily change, however. The estimator pricing construction work will always price without the addition of VAT, and all prices in the bills of quantities are similarly without the addition of VAT.

Materials and goods unfixed or off-site (16)

Practical completion and defects liability (17)

Whether or not the contractor declares rates here, the contractor must cover for rectifying defects during the defects liability (or guarantee) period. During this period, which is usually one year, the contractor takes the risk that defects will arise due to materials or workmanship not being in accordance with the specification. In fact most contractors take a broad view of defects and will repair gaps caused by timber shrinkage or plaster cracks even if the timber and the plaster were in accordance with the specification.

An allowance on the factory example might be as follows:

Inspection	4 days at £120	£480
Craftsman	8 × 39 hours at £6.10, say	1900
Materials	say	500
		£2880

Partial possession by employer (18)

Assignment and subcontractors (19)

Injury to persons and property and indemnity to employer (20)

Injury to persons and property, contractor to insure (21)

Insurance of the works, either contractor or employer to insure (22)

These insurance clauses are dealt with together here.

The insurance implications of clause 20 can be very complex. The contractor is well advised to discuss with a broker the extent to which the indemnity can be backed by an insurance policy. However, the insurance requirements for the example factory are straightforward and can be summarized as follows.

Employer's liability insurance

All employers of workpeople are required by the Employer's Liability (Compulsory Insurance) Act 1969 to carry insurance to cover claims arising from the death or injury to employees carrying out tasks during their employment. The premium charged for this insurance is based upon the payroll of the company, and therefore the cost can be built into the all-in hourly rate. The insurance is described in more detail in Chapter 2.

Public liability insurance

The contractor is required by clause 21 to carry insurance to cover any claim arising from injury, death or damage to property other than the works caused by the negligence of the contractor's operatives. The premium for this cover is calculated in the same way as for the employer's liability insurance and is therefore incorporated into the all-in hourly rate.

Joint liability insurance

Liability insurance may be required to cover jointly both the employer and the contractor against claims by third parties for damage to property other than the works but which is not caused by negligence. An example of this may be where vibration from piling operations damages a neighbouring structure. The requirement for this insurance will be detailed in the appendix to the contract, and the contractor will usually obtain quotations from a broker dealing with Lloyd's.

Contractor to insure works

Where the contractor is required to insure the works and site materials against all risks, it is necessary for the contractor to first determine the total amount to be insured. For the factory example this will be as follows:

Contract value	£1,000,000
Allowance for increased costs during construction (see later)	100,000
Professional fees at 15%	165,000
Total to be insured	1,265,000
Premium at say £2 per £1000	£2,530

Date of possession, completion and postponement (23)

Damages for non-completion (24)

This is generally a sum per week payable by the contractor to the employer for every week that the building remains incomplete.

Extension of time (15)

The extension of time clause in a contract is important because it determines where loss will fall. If the contract is delayed by actions of the architect or the employer then the contractor is allowed extra time in which to complete, and the damages described above are not payable. The contractor is also able to recover *inter alia* the cost of running the site for the extended period.

If the contractor delays the work by bad management then the contractor cannot claim either the cost of running the site for the extended period or the extension of time which alleviates the payment of damages.

If the contractor is delayed by a matter listed in the contract as being neither party's fault, then the contractor is entitled to an extension of time so that damages are not payable, but is required to cover the contractor's own cost in the tender. This is a risk item which can be calculated by asking the question: 'What is the percentage probability that this event will delay the contract by one week?' If the answer is 100 per cent then the question has to be asked again with an increased time, i.e.: 'What is the percentage probability that this event will delay the contract by two weeks?' For the example factory the risk calculation would be as follows:

What is the percentage probability that the following events will delay the contract by one week?

Bad weather	40%
Strikes	10%
Non-availability of labour	5%
Total probability of a one week delay	55%
Cost of preliminaries for one week	£5000
Risk cost = 0.55 × £5000	£2750

Loss and expense (26)

Determination by employer (27)

Determination by contractor (28)

Determination by employer or contractor (28A)

Works by employer (29)

Certificates and payments (30)

The conditions contain a provision which allows the employer to retain a percentage of each payment made to the contractor. This retention fund accumulates during the contract and is paid to the contractor, half at the point where the employer takes possession of the building and half following the satisfactory rectification of defects at the end of the defects liability period.

The effect of this retention is that the contractor is without the use of the retention money for the period of the contract plus the defects liability period. If, for the factory example, this were to be funded out of bank borrowing, then to the tender sum should be added 3 per cent retention of £1,000,000, i.e. £30,000.

As the retention fund accumulates during the contract from £0 to £30,000, it is assumed that the amount to be funded is £15,000 for one year. For the defects liability period half of the total retention fund is kept by the employer, so that £15,000 is to be funded for a further year. The cost to the contractor where bank borrowing attracts interest of 15 per cent is £15,000$(1.15^2 - 1) = $£4838.

Finance (No. 2) Act 1975: statutory tax deduction scheme (31)

Outbreak of hostilities (32)

War damage (33)

Antiquities (34)

Nominated subcontractors (35)

Nominated suppliers (36)

Fluctuations (37)

Clauses 38, 39 and 40 give the conditions under which the contractor can recover the increased cost of labour, material, plant, fuel and tax changes during the currency of the contract. The clause which applies will be indicated in the appendix to the contract.

Contribution, levy and tax fluctuations (38)

Under this clause, often called the fixed price option, the contractor is to include in the tender an allowance to cover the increasing cost of resources. The exception is for those increases resulting from changes in legislation, which are obviously unpredictable. For example, if the contractor gives his operatives a wage increase then this is not recoverable from the employer. If, on the other hand, national insurance contributions increase then the contractor will be able to recover the total cost of the changes from the employer.

The contractor is therefore required under this clause to predict the rate of inflation in terms of building costs. The contractor is not liable for the increases in cost to be carried by nominated subcontractors, and therefore the value of this work may be deducted. Domestic subcontracts are often placed during the contract period and therefore an increased cost allowance should be included for these.

In the example factory project the addition is found as follows. Assume there is 10 per cent inflation over one year, and therefore an average of 5 per cent increase in the building cost. Then

Contract value	£1,000,000
Less nominated subcontractors and suppliers	200,000
	800,000
Addition to contract 5%	£40,000

Labour and materials cost and tax fluctuations (39)

Under this option the employer elects to take the risk of cost increases, and will reimburse the contractor based upon the proved difference between the quoted price and the invoiced price. Labour rate increases are based upon those incorporated into the Working Rules agreement. The contractor has only to cover non-recoverable increases, which are basically plant and fuel. In the factory example these are insignificant.

Use of price adjustment formulas (4)

Under this option the employer elects to take the risk of cost increases, and will reimburse the contractor based upon the formula method developed by the National Economic Development Office and now maintained by the Property Services Agency. Under this method all increases are covered and the contractor has no need to make an addition.

Settlement of disputes: arbitration (41)

Appendix

The final section under clause A20 is the appendix to the contract. The appendix gives the start and completion dates, the amount of damages payable per week, and so on.

One item of significant importance is the base date. The appendix will state the date on which the tender has been prepared for all contractual purposes. This is the date on which the contractor was deemed to be aware of any announced price increases which are to be included in the tender amount. For example, if the base date was set at 4 June 1990 then the contractor is deemed to have included any price increases announced before this date. Therefore if an increase in the wage rate was announced on 3 June its effect is deemed to have been included in the tender. If it was announced on 5 June its effect is deemed not to have been included in the tender. The base date is normally set for ten days before the date for submission of tenders so that the contractor has time for any last minute alterations.

Employer's requirements: quality standards/control (A33)

Where the design team has specifically requested, the contractor is required to obtain from suppliers samples of materials which will be retained on the site for the purpose of comparison with the materials delivered. Where this involves the contractor in expense, the cost should be added to the tender sum under this item.

Concrete test cubes may be required to be made from each load of ready mixed concrete delivered to site. For the factory example this would involve the following calculation:

Approximate total volume of *in situ* concrete: $560 \, m^3$
Delivered by $10 \, m^3$ truck; two cubes from each mix
Number of cubes: $(560/10) \times 2 = 112$
Delivered to and tested by local concrete testing station at £20 per cube
Total cost: $112 \times £20 = £2240$

Employer's requirements: security/safety/protection (A34)

Site security is generally described in the bills of quantities by an item similar to the following:

> Adequately safeguard the site, the works, products, materials and plant from damage and theft. Take all reasonable precautions to prevent unauthorized access to the site.

In the factory example this may be undertaken by constructing the fencing described in the bills of quantities at the commencement of the contract and also additional fencing to create a compound for site offices and plant. The provision of the site sign board and the fixing of signs provided by the various consultants is also taken here.

Allowance for repairs to fencing	£800
Sign board	200
Compound fencing comprising chain link fence on	
concrete posts with cranked top and barbed wire:	
100 m at £40 per metre	4,000
Double gates	200
Alarm system	400
One year contract with security company	8,000
	£13,600

Site safety is a combination of safe working practices, education of operatives and monitoring. These functions are the responsibility of the site management and the safety officer whose costs are taken under clause A40.

The item for protection in the preliminaries section of bills of quantities prepared in accordance with SMM7 covers all work sections and is covered by a general clause:

> Adequately protect all types of work and all parts of the works, including work carried out by others, throughout the contract.

Unless the contract contains high quality expensive finishes which require specific protection, this clause is a risk item and is not normally priced by contractors.

Table 4.1 *On-site supervision*

Contract period: 52 weeks

Staff	Number staff*	Number cars*	Salary (£/week)	Allowances (£/week)	Total (£/week)
Agent	1	1	450	100	550
Subagent					
General foreman	1		300	150	450
Section foreman					
Trades foreman	1		250	150	400
Engineer	0.1	0.1	600	100	70
Engineer's asst	0.1	0.1	400	100	50
Chainman	0.1		200	75	28
Quantity surveyor	0.2	0.2	650	150	160
QS asst					
Office manager					
Wages clerk					
Timekeeper					
Cost clerk	1		300	150	450
Typist					
Safety officer	0.1	0.1	400	100	50
Plant manager					
Foreman fitter	—	—			—
Total per week	4.6	1.5			2,208
Cost of cars		1.5 × £100			150
Petty cash/expenses	4.6 × £50				230
Stationery	4.6 × £15				69
First aid/laundry	4.6 × £5				23
					2,680
					× 52
Total cost of supervision					£139,360

*Staff/car numbers: 1 day/week = 0.2, 1 week/year = 0.02.

Contractor's cost items: management and staff (A40)

The estimator in pricing this item will make use of a form similar to that in Table 4.1.

Contractor's cost items: site accommodation (A41)

This item comprises the accommodation required by the contractor in addition to that described in the bills of quantities for the use of the architect or contract administrator and/or the clerk of works. The cost of site accommodation includes the two components of weekly hire plus the cost of set-up and take-down. For the factory example the cost is as follows:

Weekly hire

Staff offices: 1 unit 15 m × 3.6 m at	£75/week	£75
Clerk of works office: 1 unit 5 m × 3.6 m at	£35/week	35
Operatives mess room: 1 unit 15 m × 3.6 m at	£75/week	75
Materials store: 1 hut 9 m × 6 m at	£30/week	30
Portaloo 3 cubicles, 3 urinals: 1 nr at	£35/week	35
Total weekly hire (includes furniture)		£250

Total costs

Accommodation for 52 weeks at £250	£13,000
Transportation to and from site, and set-up and take-down	400
Connection of services: 20 hours at £6.10	122
Total cost	£13,522

Contractor's cost items: services and facilities (A42)

The services and facilities item is to cover the cost of power, water, telephone, rates etc. These are listed and costed as follows:

Electricity

Connection charge	£100.00	
Wiring (included in office unit):		
Electrician 5 hours at £6.10	30.50	
Materials	25.00	
Cost of electricity 52 at £10	520.00	£675.50

Water

Temporary connection	£75.00	
Water charges at £0.02 per m^3	400.00	475.00

Telephone

Connection (site line)	£100.00	
Connection (clerk of works line)	100.00	
Fax/answering machine	500.00	
Call/rental charges (site) at £200/qtr	800.00	
Ditto clerk of works (provisional sum)	500.00	2,000.00

Local authority rates

On-site accommodation	1,000.00

Fire precautions

Extinguisher maintenance	80.00

Rubbish disposal

One skip per week at £35	1,820.00

Cleaning

Offices etc. 52 × 5 days, 2 hours/day at £5.30	£2,756.00	
Site 52 × 5 days, 2 hours/day at £5.30	2,756.00	
Roads (as necessary) say	400.00	
Professional cleaning on completion	1,000.00	6,912.00

Drying out
Dependent upon the time of year, the work of wet trades may need to be dried either by heating or by dehumidification. The example project will not need this.

Small plant and tools
This covers small plant which the contractor owns as opposed to hired plant. It includes such items as small mixers, 110 volt transformers, power hand tools, picks, shovels etc. The cost is based upon a percentage of the project cost, say 0.15 per cent of £1,000,000.   1,500.00

General attendance on nominated subcontractors
The bills of quantities measured in accordance with SMM7 will contain this general item for all nominated subcontractors. It allows the contractor to reflect the cost of providing messing and other facilities for the operatives of nominated subcontractors. In the pricing of the example contract this cost has been included in item A41 by an extra allowance in the sizing of facilities.

Site transportation
It is normal to make an allowance for a site van. This will be an all-purpose 1 tonne van for sundry collections and other errands. It will be driven by an employee and may be in the overall care of the general foreman.

1 nr 1 tonne van at £85/week	4,420.00

Contractor's cost items: mechanical plant (A43)

This item allows the contractor to price for large items of plant, the cost of which may not be totally related to one work section. A tower crane, for example, would be used for lifting skips of *in situ* concrete, steel sections, precast units etc. The cost of the tower crane would be difficult to distribute amongst these items and therefore its cost is better reflected here.

Also, in situations where work is marked as provisional, the contractor may wish to price the plant associated with that work under this section. For example, the bills of quantities may indicate piling comprising 100 driven precast concrete shell piles provisionally 15 m long. If the project demanded two piling rigs for three weeks at a cost of £3500 per week each, then the cost may be included as a lump sum of £21,000 in this section or alternatively as £14 per metre of piling.

The problem with the latter alternative is that the period that the rigs are required on site is related as much to the number of piles to be driven as to their total length. If each pile were driven 14 m instead of the 15 m anticipated, this would be unlikely to have any effect upon the total time that the rigs were required on site. However, had the pricing been based upon £14 per metre of piling the contractor would have received £1400 less income.

Contractor's cost items: temporary works (A44)

This item is provided in the bills of quantities for the pricing of any temporary works which are not to be included as an element of the bill rate. In the example factory, scaffolding is to be considered as an item under this heading. Scaffolding is most commonly provided by a specialist scaffolding company, which will quote for the erection and dismantling of scaffolding plus a hire rate per week.

It is assumed, for the purposes of the example, that scaffolding will be provided by the contractor for the brick office structure to be constructed within the framed building. From the information given in the description of the works it is assumed that a free standing scaffold will be required 48 m long × 4.5 m high × 1.2 m wide.

The names given to the various scaffolding components are as follows:

Standards   Vertical tubes 1.2 m apart, at 2 m centres along the face and back of the scaffold.
Ledgers   Horizontal tubes running the length of each side of the scaffold, at 1.5 m vertical centres.
Putlogs   Horizontal tubes across the width of the scaffold, at 2 m centres horizontally and 1.5 m centres vertically.
Braces   Diagonal tubes across the face and back of the scaffold.
Boards   Decking and toe boards.

The costs are calculated as follows:

Quantities
Standards: $[(48/2)+1] \times 2 = 50 \times 4.5$ m high	225 m
Ledgers: $(4.5/1.5) \times 2 \times 48$ m long	288 m
Putlogs: $[(48/2)+1] \times (4.5/1.5) \times 1.5$ m long	113 m
Braces: say	72 m
Total tube	698 m
Fittings: say 4 per putlog $(300)+60$	360

Boards (225 mm wide): 5 boards + toe board $= 6 \times 48$ m $= 288$ m

Erect and dismantle time
Tube:	698 m at 0.1 hours/metre	70 hours
Fittings:	360 at 0.15 hours each	54 hours
Boards:	288 m at 0.06 hours/metre	17 hours
		141 hours

Cost: erection and dismantle plus hire six weeks

Labour:	141 hours at £6.10 per hour	£860.10
Hire:	Tube 698 m at £0.02/m per week × 6 weeks	83.76
	Fittings 360 at £0.015 each per week × 6 weeks	32.40
	Boards 288 m at £0.05/m per week × 6 weeks	86.40
Transport:	say 5.5 tonnes at £30 inc. loading and unloading	165.00
Cost of scaffolding		£1227.66

Nominated subcontractors (A51)

SMM7 requires that all nominated subcontract work is described in the preliminaries section. The estimator is required to price two items in connection with this work, namely profit and special attendance. An example is as follows:

Prime cost Allow the prime cost sum of £120,000 for steelwork to be fabricated, supplied and erected by a nominated subcontractor	£120,000

Profit   Add for profit

5 per cent   6,000

This item allows the estimator to add for profit at a different rate to the profit on builder's work. If the profit is priced as a percentage it will be adjusted proportionally to the adjustment of the prime cost sum.

Special attendance   Add for special attendance: lorry mounted 7 tonne capacity telescopic crane and operator for 10 days during the construction of the steel frame. This facility is to be provided by the contractor for use by the nominated subcontractor

The cost which the estimator will include here is 12 day hire at £250 per day   3,000

Total for A51   £129,000

Provisional work items (A54)

SMM7 sets out rules for provisional sums, which are either

● Defined, in which case the estimator should consider whether there is a preliminary cost implication in the definition and include any cost here; or
● Undefined, in which case no additional cost is included in the preliminary bill. All preliminary costs associated with the expenditure of an undefined provisional sum are to be included with the valuation of work when the work is carried out.

Dayworks (A55)

Under this item the estimator is required to enter the percentage additions to the allowances made for labour, plant and material. The percentage additions are to cover the contractor's costs over and above those defined in the Definition of Prime Cost of Daywork carried out under a Building Contract, published by the RICS and the BEC.

It should be noted that under the conditions of JCT80, daywork may only be used to value work, instructed by the architect or contract administrator as a variation, where the work may not be measured.

Collection of example costs

The value of the preliminaries for the example factory are summarized in Table 4.2.

Table 4.2 *Preliminaries for factory example*

Practical completion and defects liability		£2,880.00
Insurance of the works		2,530.00
Extension of time allowance		2,750.00
3% retention financing charges		4,838.00
Fluctuations allowance		40,000.00
Concrete test cubes		2,240.00
Site security		13,600,00
Management and staff		139,360.00
Site accommodation		13,522.00
Services and facilities:		
Electricity	675.50	
Water	475.00	
Telephone	2000.00	
Local authority rates	1000.00	
Fire precautions	80.00	
Rubbish disposal	1820.00	
Cleaning	6912.00	
Small plant and tools	1500.00	
Site transportation	4420.00	18,882.50
Cost of scaffolding		1,227.66
Nominated subcontractors		129,000.00
Total cost of preliminaries		£370,830.16

5

Excavation and filling

SMM7 Section D2

Site preparation

This section is concerned with the removal of trees, tree stumps, bushes, scrub, undergrowth and hedges. A tree or tree stump is defined in SMM7 as having a trunk of at least 600 mm girth at 1 m above ground level or at the top of the stump. The assumption can only be made that anything smaller is a bush.

Guidelines for estimating for the removal of trees and stumps might include the following:

Trees   Time for squad to fell; hire of chain saws; fuel and oil; sharpening; axes, wedges, pince bars and mauls; protective clothing; time to cut up and burn the slash; disposal of the main limbs and trunks to a tip or to a firewood merchant.
Stumps   Time to excavate around larger stumps by hand, or by machine if numbers or size warrant; men excavating or hire of machine and operative; disposal of stumps to tip (firewood merchants seldom want to know about stumps).

Bushes, scrub and undergrowth can be removed by hand using chain saws or scrub saws and burning the residue. Machines can also be used to 'bulldoze' the scrub off the ground into piles for burning. Once again the choice depends on the size of the site to be cleared and the density of the growth. Burning may not be allowed – in smokeless zones or because of fire hazards nearby – and disposal to a tip may be the only solution.

Example 5.1

Felling a tree, girth at 1 m above ground level say 2 m, burn slash, and remove trunk and main limbs off site.	Nr

The only way to price the item is first of all to look at the quantity involved, and from this and the item description then to assess the likely squad size and what is required by way of plant. An isolated tree or group of up to say 10 trees can be dealt with as follows:

Felling 5 trees, cutting up, burning, loading onto lorry:

3 labourers for 2 days of 8 hours at £5.30	£254.40
Hire of chain saw, nr 2, for 2 days at £5.00/day	20.00
Lorry with MOL and driver for 2 days at £35.00/day	70.00
Charges at tip: 3 loads at £5.00/load	15.00
	359.40
Profit and oncost 20%	71.88
	431.28

Rate per tree   £86.26

A lorry with mechanical off-load (MOL) has a hydraulic crane mounted on the body. The crane can be fitted with a variety of devices for lifting different kinds of loads, e.g. pallets, round logs, slings for beams, and hooks.

Example 5.2

Remove stump and root of tree, girth 2 m at 1 m above ground level, including refilling hole with soil and disposal of stump off the site.   Nr 5

There are two ways in which stumps can be removed: the first is to dig them out and the second is to pull them out. The latter is cheaper but really requires the stump to be left with a good length of trunk intact. In this instance a powerful machine, probably tracked, is attached to the top of the stump with heavy chain. The machine literally pulls the stump out, exerting considerable leverage by virtue of the extended piece of trunk. Of course one cannot assume that, just because the tree stumps are left with a length of trunk, they can be pulled out economically. There has to be a tracked excavator on the site doing other things! The rate for stump removal would be very high if a bulldozer was to be brought onto site to take out five tree stumps such as we have described.

Excavating stumps is most easily accomplished with a backhoe on a machine such as a JCB. Having weakened the hold in the soil, the JCB might be able to pull the stump out, or the main roots may have to be cut before the machine can dislodge the stump. Excavation is treated as follows (assuming that there is an excavator on site):

Root spread approximately	1.50 m radius
Root depth approximately	1.25 m
Volume of excavation approximately	8.84 m^3

Excavation for 5 stumps: 44.2 m^3 at 10 m^3/hour at £35.00/hour inc. driver	£154.70
Backfill excavation: 44.2 m^3 at 15 m^3/hour at £35.00/hour inc. driver	103.13
Loading onto lorry and disposal: lorry time included in Example 5.1	
Charges at tip: say 5 loads at £5.00/load	25.00
	282.83
Cost per stump	56.57
Profit and oncost 20%	11.31
Rate per stump	£67.88

Lifting turf for preservation

Natural sites seldom have turf worth preserving, unless they happen to be one of the rare pieces of ground with some special soil where the turf can have an almost lawn-like quality. Turf for preservation is most likely to be encountered where the ground has been previously cultivated and is still in reasonable condition.

If the area is large, it might be worth considering obtaining quotations from a specialist landscape gardener to bring in a machine to lift and roll the turf. Lifting turf by hand is a skilled operation using a special spade and should not be undertaken by just anyone. The thickness of the turf is critical, as is the rolling, or the stacking of flat turves. If the estimator is in any doubt as to the feasibility of his own men coping with the work, then again a reliable landscape gardener should be asked to quote for the work.

Excavation: general remarks

The rules of measurement for SMM7 Section D2 have been drafted with machine excavation in mind as the principal method. Any excavation which the estimator judges can be carried out only or better by hand must be priced within the constraints of these rules. The implications of this for both the estimator and the person preparing the bill of quantities are discussed later in the chapter.

Excavation is measured in eight classes or types of excavation:

1 Top soil for preservation
2 To reduce levels
3 Basements and the like
4 Pits (give total number)
5 Trenches up to 300 mm wide
6 Trenches over 300 mm wide
7 For pile caps and ground beams between caps
8 Benching sloping ground to receive filling.

For the first class, state the average depth. For the remaining classes, give the maximum depth in stages: up to 250 mm deep; up to 1 m deep; up to 2 m deep; and so on in stages of 2 m. In addition, a starting level for the excavation should be provided where the particular class of excavation commences beyond a depth of 250 mm below existing ground level. The subclassification into depth stages with a given starting level is a blanket requirement of SMM7.

The class 'To reduce levels' has been retained from SMM6 and is assumed to be the same, i.e. the excavation takes place *above* the starting level. The blanket rule that a starting level is stated when the excavation starts beyond 250 mm below the ground level obviates any unnecessary provision of a starting level for reduce level.

The last class 'benching sloping ground to receive filling', is the only new category of excavation. It relates to the practice of excavating a permanent unsupported side slope down which there are horizontal steps which, when filled over with imported material, prevent that material slumping to the bottom of the excavation. Figure 5.1 illustrates the principle. Again the necessity to state starting levels and give depth stages can be questioned. If the excavation is a 'hole in the ground' with permanent side slopes stepped and filled over, then there is a starting level and a depth to consider. However, the most common occurrence of this type of work is in large cuttings such as are commonly encountered in motorway construction, where the excavation is entirely *above* the starting level and this last is the original ground level!

The previous two paragraphs have been critical of SMM7. Whether these criticisms are justified or not, they serve to illustrate that a quantity surveyor will interpret these rules and present the bill for pricing in the firm belief that his item descriptions and all the other supporting documentation really do represent the work properly. The estimator must be certain that he is pricing the intended work.

There is no requirement in SMM7 to state in the bills of quantities whether or not the work is to be done by machine or by hand. However, the Measurement Code published with SMM7 does state that 'Information provided in accompanying drawings or in descriptions should identify those circumstances where it may be difficult or impractical to carry out excavation by

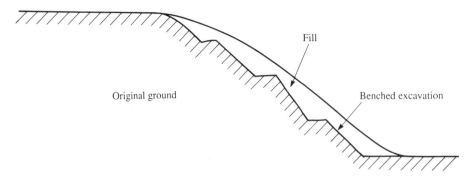

Figure 5.1   *Benching sloping ground to receive filling*

mechanical means.' The final judgement must be made solely by the estimator. To enable him to make it, the bill should clearly state whether the work is inside an existing building, under water (and whether this is river, spring or tidal) or in compressed air (SMM7, General Rule 7). A study of the drawings and the preambles will elicit further information, as will a visit to the site. However, there are further considerations which are known to the estimator and depend on his experience and the technical capabilities of his company. These are all discussed further below.

Hand excavation

Not all excavation is possible using machinery. Nor is there one type of machine which is suitable for all classes of machine excavation.

For example, excavation of foundation trenches using a backhoe on a wheeled tractor unit will not give accurate alignment of the trench, and the teeth on the bucket will disturb the bearing surface of the trench. Further, it is not possible to dig to a precise level and depth with sufficient accuracy. The backhoe will therefore take out the bulk of the excavated material – say 90 per cent – and the trench will be trimmed to level for the foundation by hand. Hand trimming to width may also be required where there is extensive timbering.

Hand excavation may also be necessary for small isolated pits for pads and manholes. excavations for foundations to entrance steps, and so on, either because it is uneconomical to bring a machine to the work or because access is restricted. In the latter instance one could also consider work inside an existing building or in a back garden or enclosed yard or area.

On the subject of trimming excavations by hand, it should be noted that no matter how the excavation is done, the smaller and more confined types of excavation have a higher proportion of trimming to volume of excavated material. Generally in the past this has been taken into account by using lower outputs per hour when calculating the overall rate, which is based purely on volume. However, SMM7 and previous editions have included provision for the preparation of excavations to receive concrete foundations.

Depth and throw

Excavating by hand – even only the trimming of machine excavated trenches, basements or whatever – brings with it the problem of reconciling the depth stages in SMM7, which are set to suit machines, with the lift and throw achievable by a man with a shovel. The norm for hole depth for man and shovel is 5 ft, which is translated loosely into 1.5 m. This is generally held to be shoulder height. Figure 5.2 and the following paragraphs explain the reduction in outputs per hour with increasing depth.

Refer first to situation A. The man in position 1 has dug out the earth and has piled it at the side of the excavation. As he has gone down, the pile of spoil at the side has risen, thus increasing further the required throw. It is for this latter reason as much as depth that a limit of 5 ft was set for depth.

Refer now to situation B. The man in position 1 continues to dig down but can no longer throw the spoil clear of the excavation and maintain a reasonable output. Reasonable output is the important point at this transition between depth stages. Of course he could throw clear if the excavation were to stop at say 1.65 m, but the effort required to clear the spoil from the excavation over the growing pile at the surface would slow him down, and that would be unacceptable were the digging to continue for even greater depths. So to keep output up, the man erects a platform or staging 1.5 m down (position 2) and puts the spoil on the platform as he continues to dig down to 3 m. Either he or a second man works off the platform shovelling the now loosened spoil up on to the side of the excavation. We have already stated that the pile of spoil at the side of the trench was making it difficult for the man excavating up to 1.5 m down, so it must also be increasingly difficult for the man shovelling from the platform. What we require to have is the loose spoil cleared back from the side of the excavation (position 3). Again this can be the one man or it could be a second man alternating between staging and starting level.

Situation C shows the saga continuing down to 4.5 m with a second staging, and the need to man both stagings and the surface to clear away.

The man at position 3 can of course do a number of operations. He can merely shovel the spoil back to make room for more and make ready for backfilling in the case of pipe trenches. He might also segregate the spoil according to the different strata being encountered, some being suitable for backfill round pipes; vegetable soil might be kept separate, and the remainder used as general backfill after a certain level. He might load everything into wheelbarrows, dumper trucks, skips or hutches.

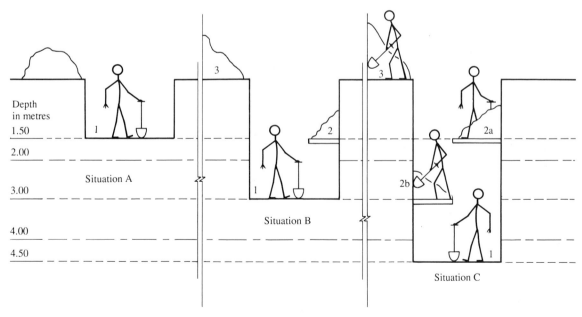

Figure 5.2   *'Throw' in manual excavations: Table 5.1 gives the times taken to get out 1 m³ spoil*

Table 5.1   *Hours to excavate 1 m³ by hand*

Depth stage (m)	To excavate and get out	To throw one stage	To clear sides	Total
<1.5	2.4	0.0	0.0	2.4
1.50 to 3.0	2.4	1.4	1.4	5.2
3.0 to 4.5	2.4	2×1.4	1.4	6.6
4.5 to 6.0	2.4	3×1.4	1.4	8.0

Table 5.2   *Multipliers for difficulty due to nature of ground and type of hand excavation*

Soil type	Multiplier ($\times$ hours/m³)	Excavation type	Comment
Vegetable soil	1.11–1.33		Because of small volume for a given area
Sand	1.25–1.52	Reduced level, basement, trench, pit	Depending on degree of compaction and moisture content, a shovel will not hold as much sand as soil
Ordinary	1.00	Reduced level, basement, trench	Any soil capable of being removed with a shovel and minimal use of the pick
	1.11–1.25	Narrow trench,	Higher proportion of
	1.11–1.33	pit	trimming
Heavy clay	1.25–2.00	Reduced level, basement, trench, pit	Graft or pneumatic spade; trimming is almost assured with accurate use of tools
Soft rock	1.43–3.00	Reduced level, basement, trench, pit	Heavy use of pick or wedges; trimming is almost assured with accurate use of tools
Hard rock	3.00–5.00	Reduced level, basement	Wedges or compressor and pneumatic drill
	5.00–10.00	Trench, pit	Trimming difficulty and confined space

Assuming the man at position 3 is clearing back ready for backfilling, then typical outputs are as shown in Figure 5.2. The man getting material out from the bottom takes almost twice as long to handle one cubic metre as those shovelling off the stagings or clearing back at the surface. No matter how many men are required, what the squad size is or what the disposition of the workers can be, the times taken to get out one cubic metre of spoil can be calculated as shown in Table 5.1. The outputs used in Table 5.1 are averaged from Tables 5.3 and 5.4, and assume ordinary soil conditions as defined in Table 5.2.

The hard part about all of this is to reconcile these ergonomically selected depth stages with those given for machine excavation in SMM7. Figure 5.2 has these depths shown as dashed lines across the illustration.

There follow two examples of rates for hand excavation, using the outputs calculated in Table 5.1.

Example 5.3

Excavate trench starting at ground level and not exceeding 2 m maximum depth.	m³

Using data from Table 5.1, and assuming excavation dimensions to be 1 m wide, then for length of 1 m on plan:

First throw stage 1.5 m deep:
  1.5 m³ at 2.4 hours/m³ at £5.30/hour   £19.08
Second throw stage only 0.5 m deep:
  0.5 m³ at 5.2 hours/m³ at £5.30/hour   13.78
Cost to excavate 2 m³   32.86

Cost per m³   16.43
Profit and oncost 20%   3.29
Rate per m³   £19.72

Example 5.4

Excavate trench starting at ground level and not exceeding 4 m³ maximum depth.	m³

Using data from Table 5.1, and assuming excavation dimensions to be 1 m by 1 m on plan:

First throw stage 1.5 m deep:
  1.5 m³ at 2.4 hours/m³ at £5.30/hour   £19.08
Second throw stage 1.5 m deep:
  1.5 m³ at 5.2 hours/m³ at £5.30/hour   41.34
Third throw stage only 1 m deep:
  1 m³ at 6.6 hours/m³ at £5.30/hour   34.98
Cost to excavate 4 m³   95.40

Cost per m³   23.85
Profit and oncost 20%   4.77
Rate per m³   £28.62

In both these examples, the staging necessary could be priced in with these rates or included with the items for earthworks support.

Soil conditions

To complicate matters further, the estimator will also be required to consider that the material

Table 5.3   *Range of outputs for various classes of hand excavation in ordinary soil (hours/m^3)*

Oversite <250 m	1.25–1.50
Basement and the like	2.00–2.25
Trenches (depending on width)	2.00–2.75
Pits (depending on size)	3.00–4.50
Reduce level	1.75–2.00
Benchings	1.75–2.00

Table 5.4   *Sundry outputs associated with hand excavation*

Volume outputs (hours/m^3)	
Wheel 50 m and deposit	0.60–0.80
Additional 50 m	0.60–0.80
Spread and level	0.40–0.50
Re-excavate from spoil heaps	1.25–2.00
Return fill and ram	1.25–1.75
Lift one stage	1.25–1.50
Clear from side of excavation	1.25–1.50
Clear from side of excavation and segregate from backfilling	1.40–1.80
Barrowing and filling hardcore in bulk	0.80–1.00
Barrowing and filling hardcore in beds <250 mm	1.15–1.30
Mechanical ramming bulk fill in 300 mm layers	0.40–0.60
Fix and withdraw timbering	7.00–10.00
Area outputs (hours/m^2)	
Grading/levelling trench bottoms	0.15–0.20
Mechanical ramming hardcore in beds	0.12–0.13
Spread and consolidate blinding	0.05–0.10

being excavated may vary in consistency from liquid to solid, that is from silt or sand below water level to hard rock such as whin or granite.

Obviously some materials are easier than others to remove, and the greatest differences are to be found in hand excavation. Table 5.2 gives a range of multipliers which may be applied to an output for normal soil to obtain outputs for various types of hand excavation in a range of soil types.

Soil conditions should be defined in the preambles section of the bill for excavation and filling, but the site visit should be used as check on all of these points. Indeed, if no trial pit or bore information is given the estimator might consider it prudent to obtain permission to have this done.

Outputs

Table 5.3 gives a range of outputs for various classes of hand excavation in ordinary soil. The variations in the figures are a consequence of a number of factors, the more important having been identified in Table 5.2.

Table 5.4 shows sundry outputs for various hand excavation tasks.

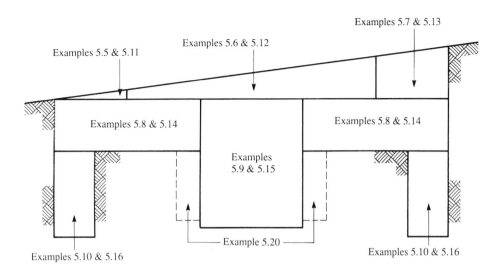

Figure 5.3 *Cross-section through a typical excavation*

Examples 5.5 to 5.10 are based on an imaginary building. A section of the earthworks is shown in Figure 5.3. The difficulty multipliers are obtained from Table 5.2, and the man hours per m³ from Tables 5.3 and 5.4.

In these examples the volume of material to be excavated is calculated for an area of unit length and unit width, i.e. 1 m × 1 m. The depth is, in the case of the reduce level digging, the average depth. For example, in Figure 5.3 the shallowest portion of reduce level digging starts at zero and rises to 0.25 m deep, giving an average depth of 0.125 m. The volume excavated is therefore 1 m × 1 m × 0.125 m = 0.125 m³. Similarly, the deepest portion of reduce level digging starts at 1 m deep and rises to 1.2 m deep, giving a total volume of 1 m × 1 m × 1.1 m = 1.1 m³ (Example 5.7).

Example 5.5

Excavate to reduce level, starting at ground level and not exceeding 0.25 m maximum depth.	m³

SMM7 has already been criticized for requiring depth stages and starting levels in items for reduce level excavation. However, the estimator knows that this work is all above the starting level and that the subsoil is a compacted sand/gravel. In this instance, the get-out part of the excavation will require barrow runs, mechanical barrows or dump trucks. Provision of barrow runs will be included in the general items in preliminaries, and the additional labour is reflected in the choice of difficulty multiplier.

Assume that we will excavate unit length and unit width at any time. Average depth (0 + 0.25)/2.

	Volume excavated	Difficulty multiplier	Man hours	Rate per hour	Cost
Max. throw from 1.5 m deep	0.125	1.11	1.5	£5.30	£1.10
Total volume excavated	0.125			Total cost	1.10

Divide total cost by total volume	8.80
Profit and oncost 20%	1.76
Rate per m^3	£10.56

Example 5.6

> Excavate to reduce level, starting at ground level and not exceeding 1 m maximum depth   m^3

Assume that we will excavate unit length and unit width at any time. Average depth $(0.25 + 1.00)/2$.

	Volume excavated	Difficulty multiplier	Man hours	Rate per hour	Cost
Max. throw from 1.5 m deep	0.625	1.2	2	£5.30	£7.95
Total volume excavated	0.625			Total cost	7.95

Divide total cost by total volume	12.72
Profit and oncost 20%	2.54
Rate per m^3	£15.26

Example 5.7

> Excavate to reduce level, starting at ground level and not exceeding 2 m   m^3

Assume that we will excavate unit length and unit width at any time. Average depth $(1.00 + 1.20)/2$.

	Volume excavated	Difficulty multiplier	Man hours	Rate per hour	Cost
Max. throw from 1.5 m deep	1.1	1.25	2	£5.30	£14.58
Total volume excavated	1.1			Total cost	14.58

Divide total cost by total volume	13.25
Profit and oncost 20%	2.65
Rate per m^3	£15.90

Note that despite the maximum depth being greater than one stage throw, there is in effect no stage throw in reduce level excavation. The result is that the rate for getting out remains the same as the previous item.

The next item, basement excavation, is 'down into' the ground, and throw may have to be taken into account if this proceeds beyond 1.5 m.

Example 5.8

Excavate basement, starting at formation level and not exceeding 1 m maximum depth	m^3

Assume that we will excavate unit length and unit width at any time.

	Volume excavated	Difficulty multiplier	Man hours	Rate per hour	Cost
Max. throw from 1.5 m deep	0.9	1.25	2.2	£5.30	£13.12
Total volume excavated	0.9			Total cost	13.12

Divide total cost by total volume	14.58
Profit and oncost 20%	2.91
Rate per m^3	£17.49

This rate assumes that the getting out will increase the hours/m^3, as it will be done on the low side of the excavations and will require the spoil to be moved over to one side of the excavation.

Example 5.9

Excavate basement starting at formation level and not exceeding 2 m maximum depth	m^3

Assume that we will excavate unit length and unit width at any time.

	Volume excavated	Difficulty multiplier	Man hours	Rate per hour	Cost
Max. throw from 1.5 m deep	1.5	1.3	2.2	£5.30	£22.74
Max. throw from 2.0 m deep	0.4	1.35	2.2	£5.30	6.30
Throw one stage 0.4 m^3		1	1.4	£5.30	2.97
Total volume excavated	1.9			Total cost	32.01

Divide total cost by total volume	16.84
Profit and oncost 20%	3.37
Rate per m^3	£20.21

Reference to Figure 5.3 will show that although the depths are all within the 1.5 m throw stage, the fact that a deeper separate excavation exists within the shallower one makes a throw a necessity. The volume thrown is *not* included in total volume.

Example 5.10

Excavate trench starting at bottom of basement excavation and not exceeding 1 m maximum depth	m^3

Without a definitive starting level, this excavation could start at one of two maximum depths below formation level: 1 m or 2 m. However, reference to Figure 5.3 will show that the trenches start at the bottom of the more shallow basement excavation. A throw still has to be added, as the initial spoil will be excavated onto the bottom of the basement excavation.

Assume that we will excavate unit length and unit width at any time.

	Volume excavated	Difficulty multiplier	Man hours	Rate per hour	Cost
Max. throw from 1.5 m deep	1	1.45	2.2	£5.30	£16.91
Throw one stage 1 m²		1	1.4	£5.30	7.42
Total volume excavated	1			Total cost	24.33

Divide total cost by total volume	24.33
Profit and oncost 20%	4.87
Rate per m^3	£29.20

Machine excavation

The ubiquitous tractor with backhoe is not the only machine available. Giant earth moving and excavating equipment is capable of moving mountains. Miniature machines can be driven through a 1 m gap to give machine capabilities in back gardens or even inside buildings themselves.

Here we must repeat the earlier caveat regarding the suitability of a particular machine for the type of excavation being tackled. Texts on contractors' plant treat the subject of choice of appropriate equipment in some depth.

With regard to soil conditions, machine excavation is not subject to the same variation in output as hand work, providing the correct type of plant can be procured. The output multipliers in Table 5.5 give some guide as to variations in performance in different soil conditions.

Table 5.6 shows outputs for a particular machine for different work and conditions. The soil classifications given in the table, which clearly influence the output of the machine, seem to obviate the necessity to use the difficulty multipliers from Table 5.5. This is only partly true. The differences across Table 5.6 are more to do with how much of the spoil is retained in the bucket at each cycle rather than how the machine can cope with penetrating the soil. Unless the work is in soft or hard rock, penetration is not a great problem for the larger modern excavator

Table 5.5 *Multipliers for difficuty of machine excavation due to nature of ground*

Soil type	Multiplier ($\times$ output per hour/m^3)
Vegetable soil, sand, ordinary soil	1.0
Heavy clay	1.33
Soft rock	2.00
Hard rock	5.0*

* Low output due to general inability of machines to do the breaking up, which requires pneumatic drills or explosives.

Table 5.6 *Outputs for a tractor with front shovel (1.24 m^3) and backhoe (0.06–0.28 m^3): loading of spoil is straight into dumpers or tipping lorries alongside excavation.*

Excavation	Bucket size (m^3)	Loose granular (small) (m^3/hour)	(hour/m^3)	Compact organic soil (m^3/hour)	(hour/m^3)	Soft cohesive (m^3/hour)	(hour/m^3)	Firm cohesive and compact granular (large) (m^3/hour)	(hour/m^3)
Trenches	0.60	2.5	0.40	2	0.50	1.6	0.63	1.4	0.71
	0.12	4	0.25	3	0.33	2	0.50	1.8	0.56
	0.19	6	0.17	5	0.20	4	0.25	4	0.25
	0.28	9	0.11	8	0.13	7	0.14	6	0.17
Basements	0.06	4	0.25	3.5	0.29	2.5	0.40	2.5	0.40
	0.12	6	0.17	5	0.20	4	0.25	4	0.25
	0.19	9	0.11	8	0.13	6	0.17	6	0.17
	0.28	13	0.08	12	0.08	10	0.10	9	0.11

with good reserves of power. However, the data in Table 5.5 can be useful where excavation rates for ordinary soil only are available.

Examples 5.11 to 5.16 refer again to Figure 5.3. The difficulty multipliers are obtained from Table 5.5, and the machine hours per m^3 from Table 5.6.

Example 5.11

Excavate to reduce level, starting at ground level and not exceeding 0.25 m maximum depth	m^3

Assume that we will excavate unit length and unit width at any time.

	Volume excavated	Difficulty multiplier	Machine hours	Rate per hour	Cost
	0.125	1	0.09	£10.50	£0.12
Total volume	0.125			Total cost	0.12

Divide total cost by total volume	0.96
Profit and oncost 20%	0.19
Rate per m^3	£1.15

The output in machine hours per m^3 used for the reduce level excavations have been taken as 20 per cent better than the equivalent rate for basement excavation, the reasoning being that the dumpers or tipper lorries will get alongside the machine at the face being dug.

Example 5.12

Excavate to reduce level, starting at ground level and not exceeding 1 m maximum depth	m^3

Assume that we will excavate unit length and unit width at any time.

	Volume excavated	Difficulty multiplier	Machine hours	Rate per hour	Cost
Max. throw from 1.5 m deep	0.625	1	0.09	£10.50	£0.60
Total volume excavated	0.625			Total cost	0.60

Divide total cost by total volume	0.96
Profit and oncost 20%	0.19
Rate per m^3	£1.15

Example 5.13

Excavate to reduce level, starting at ground level and not exceeding 2 m depth.	m^3

Assume that we will excavate unit length and unit width at any time.

	Volume excavated	Difficulty multiplier	Machine hours	Rate per hour	Cost
	1.1	1	0.09	£10.50	£1.04
Total volume excavated	1.1			Total cost	1.04

Divide total cost by total volume		0.95
Profit and oncost 20%		0.19
Rate per m³		£1.14

Example 5.14

> Basement excavation starting at formation level and not exceeding 1 m maximum depth.   m³

Assume that we will excavate unit length and unit width at any time.

	Volume excavated	Difficulty multiplier	Machine hours	Rate per hour	Cost
	0.9	1	0.11	£10.50	£1.04
Total volume	0.9			Total cost	1.04

Divide total cost by total volume	1.16
Profit and oncost 20%	0.23
Rate per m³	£1.39

Example 5.15

> Basement excavation starting at formation level and not exceeding 2 m maximum depth.   m³

Assume that we will excavate unit length and unit width at any time.

	Volume excavated	Difficulty multiplier	Machine hours	Rate per hour	Cost
	1.9	1	0.11	£10.50	£2.19
Total volume	1.9			Total cost	2.19

Divide total cost by total volume	1.15
Profit and oncost 20%	0.23
Rate per m³	£1.38

Example 5.16

> Excavate trench starting at bottom of basement excavation and not exceeding 1 m maximum depth.   m³

Assume that we will excavate unit length and unit width at any time.

	Volume excavated	Difficulty multiplier	Machine hours	Rate per hour	Cost
	1	1	0.17	£10.50	£1.79
Total volume	1			Total cost	1.79

Divide total cost by total volume	1.79
Profit and oncost 20%	0.36
Rate per m³	£2.15

Ground water

Excavation below ground water level is measured extra over any type of excavation, irrespective of depth.

Where the whole excavation is below ground water level, the estimator will calculate both full rates – excavation above and excavation below – and subtract the former from the latter to obtain the extra over rate.

In other cases the estimator has to decide for himself, how much of any one type of excavation will be below water level. For example, complicated excavations might include all classes of excavation – basement, trench and pit, all to different depths – and some or all might be below or partly below water level. The quantity surveyor knows but he does not tell! Outputs vary for each class of excavation and for each depth of each class, and they must do so irrespective of whether there is water in the way or not.

The factors influencing the estimator's judgement on excavation below ground water level are as follows:

● Men working in the excavations require protective clothing.
● Outputs for hand work are generally halved.
● Outputs for machine work are generally halved.
● The proportion of various classes of excavation below water table must be considered.
● Pumping, although given separately in the bill, must be realistically assessed and assumed to give optimal working conditions.

Services

Excavation adjacent to, across or under existing services is also measured extra over any type of excavation.

There may be underground or overground services such as water mains, oil pipe lines, gas pipes and electrical cables, as well as water pipes, telephone wires, district heating mains, culverts, sewers and drain pipes. All of these present a hazard. They may also be old, corroded or already damaged in some way and therefore in a delicate state.

A visit to the offices of the various utilities will determine whether any services are buried on the site. Utilities personnel are willing to spend time with detectors etc. on site to ensure that you avoid their particular service pipe or cable. A further alternative is to use something like the SUSIEPHONE service provided in Lothian Region to obtain the information.

The factors affecting the estimator's judgement on excavation near services are as follows:

● Some utilities, such as gas and electricity, will require close liaison and this can slow down the work.

- Machine use is restricted at crossings of services.
- Only hand excavation may be allowed where services are adjacent to the excavation.
- Damage to services may result in heavy penalties as well as costs for repairs or replacement.

Example 5.17

Extra over any excavation for excavating below ground water level.	m^3

We will assume that this occurs in the building shown in Figure 5.3, and affects the basement excavation to 2 m deep. The ordinary rate is given in Example 5.9 as £20.21 per m^3.

Assume that we will excavate unit length and unit width at any time.

	Volume excavated	Difficulty multiplier	Man hours	Rate per hour	Cost
To 1.5 m deep	1.5	2.6	2.2	£5.30	£45.47
To 2.0 m deep	0.5	2.7	2.2	£5.30	15.74
Throw one stage 0.5 m^3		1	1.4	£5.30	£3.71
Total volume	2.0			Total cost	64.92

Divide total cost by total volume	32.46
Deduct basic cost as Example 5.9	16.84
	15.62
Profit and oncost 20%	3.12
Extra over rate per m^3	£18.74

Breaking out

This is the term applied to the removal of material from the excavations which requires the use of equipment other than pick and shovel. These materials are defined in SMM7 as

- Rock
- Concrete
- Reinforced concrete
- Brickwork, blockwork or stonework
- Coated macadam or asphalt.

It is important to note that it is relatively easier to break out any of these materials if they have been laid down in a thin layer – say up to 300 mm thick. It is possible to come across rock in such layers, although there are usually several such layers interleaved with softer material; nevertheless these are easier to remove.

Hand tools such as pick and shovel are of little use against concrete, high quality brickwork or bituminous macadam. A compressor and pneumatic drills are required to break up these materials, whether in bulk or in layers. The hire cost of a compressor and drills or even of electrically powered hammer drills is so low that wedges and hammers are falling out of use.

Once broken up, the material may be loaded for disposal with either a hand shovel or a machine.

Outputs and multipliers for excavation in earlier tables reflect the use of appropriate mechanical means of breaking out difficult materials. Generally this means using a compressor with two drills equipped with suitable points.

It should be noted that the outputs for hand or machine loading for disposal, while being very different from each other, will not vary much from material to material. In addition, these materials will all bulk up by 33 to 50 per cent.

Breaking out with explosives is such a rare occurrence in building works, and so specialized, that it has not been included.

Example 5.18

Extra over any type of excavation for breaking out rock.	m³

We will assume that there is rock in the trenches in Figure 5.3. The ordinary rate is given in Example 5.10 as £29.20 per m³.

		Hours per m³	Rate per hour	Cost
Compressor and rock points	1 m³	7	£2.80†	£19.60
Labour on compressor and drills	3 nr	7	£5.44	114.24
Get out loose materials	1.3 m³	3.5	£5.30	24.11
Throw one stage*	1.3 m³	1.5	£5.30	10.33
Total volume	1	Total cost		168.29

Divide total cost by total volume	168.29
Deduct basic cost as Example 5.10	24.33
	143.96
Profit and oncost 20%	28.79
Extra over rate per m³	£172.75

* Note that a bulking factor of 30 per cent has been added for getting out and throwing. Note also that the output for the compressor breaking out is very low as the work is being carried out in a trench. In more open conditions the output would be 4 to 4.5 hours/m³.

† Note that the hourly rate includes fuel, oil, grease, labour in attendance, maintenance and sharpening points.

Example 5.19

Extra over any type of excavation for breaking out a layer of tarmacadam 300 mm thick.	m³

We will assume that there is an area of old tarmacadam 300 mm thick on the surface where the reduce level excavation will be carried out (Figure 5.3). There are three distinct rates for Examples 5.11, 5.12 and 5.13, which makes it difficult to decide which to use as the base rate. This would normally be done on the basis of the proportion of the bill quantities for each item. In this instance we assume that Example 5.6 gives the largest item: here the ordinary rate is given as £15.26 per m^3.

		Hours per m^3	Rate per hour	Cost
Compressor and chisels	0.3 m^3	0.33	£2.80	£0.28
Labour on drills	3 nr	0.33	£5.44	5.39
Get out loose materials	0.4 m3	2.4	£5.30	5.09
Total volume	0.3	Total cost		10.76

Divide total cost by total volume	35.87
Deduct basic cost as Example 5.6	12.72
	23.15
Profit and oncost 20%	4.63
Extra over rate per m^3	£27.78

Note that an allowance for bulking has been added for getting out.

Working space

The measurement for working space in SMM7 is now given in square metres, provided that the face of the work requiring space is less than 600 mm from the face of the excavation. There is no minimum depth below which working space is not measurable; if it is required it must be measured. At first sight this change in SMM7 might appear to be quite radical, but to the estimator it only rationalizes what he has always had to do: assess for himself exactly how much working space would be required, whether or not any was given in the bill!

All previous methods of measurement have treated working space as a theoretical allowance, where the distance from the surface worked upon to the face of the excavation was set in accordance with the depth of the particular class of excavation. Invariably all other factors, such as the nature of the soil, were ignored. Some methods of measurement did try to relate some of the difficulty factors to the measurement, but the allowance was still theoretical.

The problem then that the estimator faces is twofold. It is necessary to decide first how much space the workmen will need, and secondly how many external and internal corners there are on the plan! The first problem requires answers to the following:

● To what depth will the work be carried out?
● What kind of work is involved?
● Is the working space partly or wholly below water level and, if partly, by how much?
● Will the backfilling be of any special materials? If so, then disposal of the excavated material will be required.
● How much additional support will the excavations require?

The second problem only occurs if the building is not rectangular in plan. If it is rectangular then there are always four corners to add to the working space excavation, no matter how many indents or cutouts there are. Should the building plan be a polygon with more than four sides then careful note of the internal and external corners should be made, so that the extra excavation, disposal and support will be properly accounted for.

Having sorted these problems out, the estimator has to decide what width of excavation to allow, cube it up, price it, extend a total and divide by the bill quantity in square metres to arrive at a rate! But what about that width? This is another case for the estimator's skill and working knowledge, backed up by meticulously kept records of earlier work. Every job must be assessed on its merits.

Example 5.20

> Extra over basement excavation for working space excavation starting at bottom of shallow basement excavation and not exceeding 1 m deep.   m²

Refer to Figure 5.4 for details of what causes the need for working allowance, how much is measured and how much we must excavate and backfill. The working space excavation will start below the more shallow basement excavation and go down to the top of the concrete slab. It will therefore be within 1.5 m maximum throw depth.

Assume that we will excavate unit length and unit width at any time.

	Volume excavated	Difficulty multiplier	Man hours	Rate per hour	Cost
Max. throw from 1.5 m deep	1	1.25	2.2	£5.30	£14.58
Backfill	1	1	1.6	£5.30	8.48
Dispose of surplus*	0.2	1	0.7	£5.30	0.74
Total volume	1			Total cost	23.80

Figure 5.4   *Allowance for working space excavation*

Girth of face of protective brickwork	22.00 m
Actual depth of working space excavation	0.80 m
Area for working space given in bill	37.40 m²
Width for working space	0.60 m
Volume of digging $22.00 \times 0.80 \times 0.55$	9.68 m³
Plus corners $4 \times 0.55 \times 0.55 \times 0.8$	0.97 m³
Total excavation	10.65 m³

Excavation cost: total cost × total excavation (£23.80 × 10.65)	253.47
Additional strutting, 10 reuses, 30 m at £220.00 per 100 m	6.60
Labour setting strutting at 10 hours/m³, 0.3 m³ at £5.59/hour	16.77
	276.84
Profit and oncost 20%	55.37
	£332.21
Rate per m² of working space	£8.98

*Bulking must be disposed of.

Earthwork support

Earthwork support is measurable to all faces of excavation except the following:

● Faces not exceeding 250 mm high
● Sloping faces where the angle of inclination is less than 45° to the horizontal
● Faces which abut existing structures.

Measurement is classed in SMM7 in depth stages and in distances between opposing faces. It appears that the intention is to give gross areas of support where the total depth of excavation falls within any depth stage.

Up to 4 m between opposing faces it is possible to span across with struts of various kinds, e.g. timbers wedged in or proprietary telescopic struts of steel. Over 4 m it is usual to set up raking supports to the lower level of the excavation, although if this is below water level it may be better to use steel work for the shoring, and this will add considerably to the cost. Other special conditions must be given in the bill, i.e. where the face is curved, below water level (there does not appear to be any provision for support partly below water level), next to roadways, next to existing buildings or in unstable ground, and finally where the support is left in place.

The estimator is left to decide how to actually achieve the support, and he will take into account the following additional considerations:

● The most suitable material must be chosen, for example: solid timbers for walings, poling boards or strutting; steel sheet, plywood etc. for support; proprietary struts; spacing of all members considering the soil type; proprietary close sheeting system.
● The number of uses must be estimated, i.e. the number of times each component can be reused before being discarded. Figures are quoted from 5 to 15 uses.
● Men working in the excavations require protective clothing.
● Outputs for excavation round timbering are generally halved.
● The proportion of various classes of excavation below the water table must be considered
● Pumping, although given separately in the bill, must be realistically assessed.

Example 5.21

> Earthwork support not exceeding 1 m maximum depth, and distance
> between opposing faces not exceeding 2 m.   m²

The price for this will vary according to the nature of the soil to be supported. For
fairly firm, stable soil allow poling boards at 400 to 600 mm centres, two walings in the
height, and pairs of struts at 2 m centres.

Take a 12 m length of trench and cost it out.

Polings: 2 sides, 12 m long, set at 450 mm centres:		
nr 53 + 2 = 55, each 1 × 0.2 × 0.038 m		0.42 m³
Walings: 2 sides, 2 per side:		
nr 4, each 12 × 0.2 × 0.038 m		0.36 m³
Strutting: at 2 m centres gives 6 + 1 = 7 pairs:		
nr 14, each 2 × 0.1 × 0.1 m		0.28 m³
Volume		1.06 m³
Cost at £220 per m³		£233.20
Waste 10%		23.32
Cost timbering		256.52
Cost per use if 10 uses		25.65
Labour to unload, set up, strike on completion:		
1.06 m³ at 10 hours/m³ at £5.59		59.25
Cost for trench		84.90
Trench 12 m long and 1 m deep requires 24 m²:		
Cost for 1 m²		3.54
Profit and oncost 20%		0.71
Rate per m²		£4.25

Example 5.22

> For soil such as we proposed for our imaginary building of Figure 5.3 it
> might be necessary to put the poling boards closer and use stronger
> walings than in Example 5.21. This example illustrates the differences.   m²

Take a 12 m length of trench and cost it out.

Polings: 2 sides, 12 m long, set at 400 mm centres:		
nr 60 + 2 = 62, each 1 × 0.2 × 0.038 m		0.47 m³
Walings: 2 sides, 2 per side:		
nr 4, each 12 × 0.2 × 0.05 m		0.48 m³
Strutting: at 2 m centres gives 6 + 1 = 7 pairs:		
nr 14, each 2 × 0.1 × 0.125 m		0.35 m³
Volume		1.30 m³

Cost at £220 per m³	£286.00
Waste 10%	28.60
Cost timbering	314.60
Cost per use if 10 uses	31.46
Labour to unload, set up, strike: 1.30 m³ at 10 hours/m³ at £5.59	72.67
Cost for trench	104.13
Trench 12 m long and 1 m deep requires 24 m²:	
Cost for 1 m²	4.34
Profit and oncost 20%	0.87
Rate per m²	£5.21

Example 5.23

Earthwork support not exceeding 2 m maximum depth, and distance between opposing faces not exceeding 2 m.	m²

Assume the same loose soil as in Example 5.22.

Take a 12 m length of trench and cost it out.

Polings: 2 sides, 12 m long, set at 400 mm centres:	
nr 60 + 2 = 62, each 2 × 0.2 × 0.05 m	1.24 m³
Walings: 2 sides, 3 per side:	
nr 6, each 12 × 0.2 × 0.05 m	0.72 m³
Strutting: at 1.5 m centres gives 8 + 1 = 9 sets of 3:	
nr 27, each 2 × 0.15 × 0.15 m	1.21 m³
Volume	3.17 m³

Cost at £220 per m³	£697.40
Waste 10%	69.74
Cost timbering	767.40
Cost per use if 10 uses	76.71
Labour to unload, set up, strike: 3.17 m³ at 10 hours/m³ at £5.59	177.20
Cost for trench	253.91
Trench 12 m long and 2 m deep requires 48 m²:	
Cost for 1 m²	5.29
Profit and oncost 20%	1.06
Rate per m²	£6.35

Disposal of water

SMM7 recognizes two sources of water affecting excavations: water present in the ground, and water accumulating on the surface of the ground. The estimator is expected to give lump sum prices for dealing with either or both. The former arises when excavation takes place below the natural water table of the ground and the water wells up into the excavation. The lump sum price given in the bill by the estimator is adjusted in the final account in relation to measurements of ground water levels taken for each excavation as they are done.

The following points are valid for both classes of disposal:

- Soil type
- Topography of the site
- Methodology adopted by the contractor for the construction in general and for the disposal of ground and/or surface water in particular
- Time of year and expected weather pattern
- Records from previous contracts in the area.

Disposal of excavated material

The primary factors in disposal of excavated material are as follows:

- What type and condition is the material?
- It is suitable for backfilling?
- Is it suitable for general filling?
- Has it to be disposed of in a particular way, i.e. in spoil heaps for later use, or to a coup? Or does the contractor have the choice?

Other factors are the distance to spoil heaps or the proximity of a tip external to the site, the value of any surplus material, and whether or not the client wishes to retain a proprietary interest in it up to the point of disposal. For example, hand excavation followed by disposal into spoil heaps on site carried out using wheelbarrows might be reasonable if the wheeling distance was not too far, the ground was suitable or barrow runs could be economically provided. A maximum distance might be in the region of 150 m. Beyond this distance mechanical barrows or small dump trucks might be more suitable, bearing in mind that mechanical devices run on four wheels and need wider roadways, and that they bog down in poor soil conditions.

Wheeled or tracked plant engaged in bulk excavations such as reduced level or basement work could 'wheel' each bucket load to a disposal point, but this is only economic if the bucket is of reasonable capacity, the wheeling distance is less than 50 m and the machine can get out of the excavation to deposit the load. For distances beyond this it is considered more economical to utilize dump trucks or light tramway and hutches for on-site disposal, and tipping lorries for off-site disposal.

When considering disposal, the estimator must be aware that excavated material is measured in the bill as the volume of the void to be created, i.e. the volume of *compacted* spoil removed. As soon as it is removed it will increase in volume. This phenomenon is called *bulking* and is ignored by the quantity surveyor when measuring disposal. However, the estimator must allow in his rates for the fact that whenever he disposes of one metre cube of spoil as measured, the men on site will be handling one and a bit metres cube of spoil (see Table 5.7).

Table 5.7   *Range of increase in bulk for various soil types*

Type of soil	Rate of bulking (%)
Vegetable soil	25–30
Gravel	10–20
Sand	10–12.5
Clay	25–50
Chalk	33–50
Rock	40–50

Table 5.8   *Summary of factors affecting rates and choice of hand or machine excavation and disposal*

Type of excavation
Shallow area: vegetable soil excavation
Open excavation: reduced level, basement
Confined excavation: trench
Isolated excavation: pit

Nature of excavation
New site: unrestrained or no access for machines
Existing site or buildings: no access for machines; inside existing building

Restrictive conditions by client or local authority
Work restricted to certain hours
Routes to tip restricted
Compressors not allowed
Blasting not allowed

Nature of ground
Dry, wet, frozen
Clay, sand, gravel, rock etc.: degree of compaction, requiring pick and shovel or compressor and tools; rate of bulking
Water table: above or below; seasonal fluctuations; tidal fluctuations; pumping required

Weather
Frozen ground as above
Diversion or pumping of surface water

Volume of excavation
Too large for just hand excavation
Too little to warrant expense of machine
Mixed: bulk by machine, trim by hand, small isolated by hand
Variety of excavation, requires more than one type of machine

Disposal
Haul to tip: fees; spreading at tip; length of haul
On site: spread and level; spoil heaps re-excavated; dispose, distance of wheel

The corollary of course is that the cubic metre handled on site is not as heavy as when it was in the ground. This is important when loading lorries which have axle weight restrictions when on the public roads. For example, loading a 6 tonne payload lorry with soil having a weight of $1.5\,t/m^3$ and a bulking rate of 15 per cent means that the lorry might carry $(6/1.5)+15\% = 4.6\,m^3$. However, the soil must not be beyond the bulk capacity of the lorry body, causing it to overflow onto the roads; the authorities do not allow that either.

Matching the output of the men and/or machines digging to the means of disposal is of great importance if maximum efficiency is to be achieved. Whether disposal is by barrow or lorry, it is costly to have the excavation lagging behind the disposal or vice versa. They must match, as explained in Chapter 3 on mechanical plant.

Table 5.8 summarizes the factors for consideration in excavation and disposal.

Example 5.24

Sundry items of disposal by hand. Barrows and planking for barrow runs are included in general overhead percentages.

Wheel 50 m and deposit in heaps.		m^3
Load into barrow	1.5	
Wheel and deposit	0.7	
	2.2 hours at £5.30	£11.66
Profit and oncost 20%		2.33
Rate per m^3		£13.99

Wheel 100 m and deposit in heaps.		m^3
Load into barrow	1.5	
Wheel and deposit	0.7	
Additional 50 m	0.7	
	2.9 hours at £5.30	£15.37
Profit and oncost 20%		3.07
Rate per m^3		£18.44

Wheel 50 m and deposit, spread and level.		m^3
Load into barrow	1.5	
Wheel and deposit	0.7	
Spread and level	0.4	
	2.6 hours at £5.30	£13.78
Profit and oncost 20%		2.76
Rate per m^3		£16.54

Return, fill and ram with material from the excavations.		m^3

Excavate from spoil heap	1.5	
Return, fill and ram	1.6	
	3.1 hours at £5.30	£16.43
Profit and oncost 20%		3.29
Rate per m^3		£19.72

Note that in the last example the spoil is assumed to be at the side of the excavation where it was left immediately on excavation.

Example 5.25

A basement excavation is being dug by tracked excavator. The excavator and the dumpers will gain access to the excavations by a temporary earth ramp. We will assume that the excavation has been rated and we now wish to price the disposal to a coup two miles from the site. There is a total of 740 m^3 of excavation. Calculate the rate per m^3.

The excavator has a 0.57 m^3 bucket and is expected to operate in the region of 20 cycles per hour. Each cycle is: excavate one full bucket; off-load into dump truck; return to face of excavations. This gives an excavation rate of 11.4 m^3 per hour. The excavator will be operative for 35 hours per week.

The material is assumed to bulk up by 15 per cent on excavation to 13.11 m^3. Dump trucks must therefore be capable of handling 13.11 m^3 per hour. The excavation rate produces 1 m^3 in $60/13.11 = 4.58$ minutes. The time taken to load a 2 m^3 dump truck is then 9.15 minutes, The time taken to reach the tip at 20 mph is $60(2/20) = 6$ minutes, and the return time is another 6 minutes. The time taken to tip the load is say 3 minutes. Thus

$$\text{number dumpers required} = \frac{\text{time to load and haul}}{\text{time to load}} = \frac{9.15 + 15}{9.15} = 2.639$$

This is rounded up to three trucks. It is better to have a truck idle for a short period than to have a more expensive excavator sitting around waiting for every dump truck to return. Besides which, it is perfectly possible for dump trucks to be delayed in traffic at peak periods in the day.

Hire of dump trucks and road licence per week	£190.00
Driver at £5.30+0.29=£5.59/hour for 39 hour week	218.01
	408.01
Cost per hour over 35 hour week	11.66
Cleaning and maintenance at half hour per day of 8 hours, i.e. 1/16 of driver cost per hour	0.35
Fuel, say 2.5 litres/hour at £0.36/litre	0.90
Oil and grease, say, per hour	0.10
Insurance at 5% of cost per hour	0.58
Cost of one dumper truck	13.59
Cost of two more dumper trucks	27.18
Cost per hour	40.77

The output is related to the excavator at 13.11 m³/hour. Divide cost/hour of dump trucks by excavator output/hour	3.11
Transporting to site, own wheels, say half hour each way, divided by expected total output of 740 m³	0.06
Local dumping charge per m³	0.50
	3.67
Profit and oncost 20%	0.73
Rate per m³	£4.40

Filling to excavations

The volume of fill given in the bill should be the volume of the void to be filled. Thus, whether the material used is to come from the excavation in the first instance or to be imported to the site, the estimator must be aware that the volume of material required is much greater, for it will require compaction after placing. For example, hardcore fill decreases by 25 per cent of its original bulk after placing and compaction.

SMM7 gives two classifications of filling: filling to excavations and filling to make up levels. Filling to excavations is generally more labour intensive or more difficult by machine since it involves backfilling around the actual building works. Filling to make up levels is assumed to mean bulk fill over more extensive areas either inside the building or in the open, where it would be possible to load in with a machine or by directly tipping a truck. The latter is obviously the cheapest solution, although not strictly speaking the best construction method since any fill should be laid down in layers and compacted before the next layer is superimposed.

SMM7 also gives two thickness classifications, up to 250 mm and more than 250 mm, thus recognizing the additional labour in spreading thinner layers over wide areas.

Example 5.26

Filling exceeding 250 mm thick to make up levels with material from the excavations, well rammed in layers of 250 mm thickness.	m³

Excavation in spoil heaps	1.8	
Barrow in and fill	0.8	
Mechanical punning	0.5	
	3.1 hours at 5.30	£16.43
Mechanical punner hire inc. fuel, oil etc.: 0.5 hours at £3.00		1.50
		17.93
Profit and oncost 20%		3.59
Rate per m³		£21.52

Example 5.27

Filling not exceeding 250 mm average thickness to make up levels with hardcore of broken brick or stone, blinded with ashes.	m^3

From inspection of the drawings we ascertain the hardcore bed is 150 mm thick. Allow 6 m² per cubic metre.

Barrow in and fill hardcore 1 m³: 1.3 hours at £5.30	£6.89
Mechanical punning: 6 × 0.13 hours at £5.30	4.13
Spread and consolidate blinding: 6 × 0.7 hours at £5.30	22.26
Punner hire inc. fuel, oil etc.: 0.78 hours at £3.00	2.34
	35.62
Hardcore at £7.65/tonne, 1.6 tonne/m³: £12.24/m³	
20% loss in compaction, add 25%: £3.06/m³	15.30
Ash blinding at £5.00/tonne, 0.8 tonne/m³: £4.00/m³	
Layer 25 mm thick over 6 m² gives 0.15 m³: cost per m³ of hard core	0.60
	51.52
Profit and oncost 20%	10.30
Rate per m³	£61.82

Surface packing

According to SMM7, no item need be given for achieving a finished surface anywhere between the horizontal and a slope of 15°. Over 15° there are two classifications: battered and vertical. Battered surfaces up to the natural angle of repose of the material can be readily achieved using machines, but beyond this angle material would have to be packed or built by hand. The angle is usually in the region of 40°. Packing by hand is very time consuming.

Surface treatments

Five categories are given in SMM7:

1 Application of herbicides is self-explanatory, but it should be pointed out that men applying herbicides do require protective clothing, proper washing facilities and the means to dispose of or safely store protective clothing.
2 Compacting is subdivided into three classifications and needs no explanation.
3 Trimming the sides of cuttings or embankments is subdivided into battered, vertical or rock; the last would be natural rock and would therefore only occur in cuttings.
4 Trimming rock to a fair or exposed face.
5 Preparing subsoil for topsoil.

Some features of these have already been covered in the examples.

Example 5.28

Grade bottom of excavation to a level surface.	m^2

Grade/level: 0.08 hours at £5.30   £0.42
Profit and oncost 20%   0.08

Rate per m^2 £0.50

Example 5.29

Grade and compact surface of hardcore bed to a slope of 20 degrees.	m^2

Hand pack hardcore: 0.13 hours at £5.30   £0.69
Profit and oncost 20%   0.14

Rate per m^2 £0.83

6

Concrete work

SMM7 Sections E11, E20, E30, F31

In situ concrete

Concrete is a mixture of cement, fine aggregate, coarse aggregate and water. The strength and durability of concrete are dependent on the proportions of materials in a particular mix. The methods for specifying prescribed and designed mixes of concrete, either site mixed or ready mixed, are covered by BS 5328.

Designed mixes

Designed mixes are those for which the purchaser is responsible for specifying the performance, and the producer is responsible for selecting mix proportions that will conform with the particular performance requirements.

Prescribed mixes

Prescribed mixes are those for which the purchaser specifies the mix proportions that will produce a concrete with the required performance. There are two types of prescribed mix:

Ordinary prescribed mixes   These are specified in accordance with the requirements of BS 5328. The grade of concrete is prescribed in Table 1 of the standard, together with permitted types of cement and aggregate, aggregate size, etc.

Special prescribed mixes   These generally meet the same requirements as ordinary prescribed mixes, but mix proportions in kilograms are given for each constituent material.

Materials

Cement

The most widely used cement is ordinary Portland cement, which is manufactured to perform to B2 12. In the examples that follow we use ordinary Portland cement. There are of course other types of cements produced to satisfy particular criteria.

Aggregate

The term 'aggregate' is used to describe the gravels, crushed stone and other materials which are mixed with cement and water to make concrete. Aggregates defined as fine aggregate are natural sands, crushed rock or gravel or other materials which will pass through a 5 mm BS sieve. Coarse aggregate, such as natural gravel, crushed rock or crushed gravel, is retained in this sieve.

Water

Water for use in concrete should be reasonably free from such impurities as suspended solids, organic matter and dissolved salts, which may adversely affect the properties of the concrete.

Admixtures

No admixtures are permitted to be incorporated in ordinary prescribed mixes. Therefore the development of costs in the examples in this chapter does not include costs for admixtures. However, the estimator must be aware of and account for any such requirement in the specification for admixtures in designed or special prescribed mixes.

Weights and volumes

The unit of measurement for the main structural items of *in situ* concrete work is the cubic metre (m^3). However, constituent materials that form concrete are sold by weight. Thus we must develop the costs for such items by weight, and then convert the costs by weight to costs by volume for dry materials. Table 6.1 gives average weights per cubic metre for various materials.

Weights of coarse and fine aggregates may vary for a number of reasons, and the estimator can account for these variations where required. For example, the weight per volume depends on how wet or dry the material is, and on the source of the material.

The mass of aggregate to be used with 100 kg of cement can be obtained from Tables 1 and 2 of BS 5328.

Shrinkage in concrete mix

During the process of mixing materials to form concrete, a reduction in bulk takes place. This is due to the finer particles of sand and cement filling the voids or interstices of the coarse aggregate. In order to attain the required volume in a particular mix we have to increase the quantities of material to compensate for the reduction in bulk. This is done by calculating the cost based on the mix proportions of the concrete and adding a percentage adjustment to allow for the cost of the additional material. There are various methods of adjusting for shrinkage in concrete, as detailed below. The preference of method will vary from estimator to estimator.

Table 6.1   *Average weight per cubic metre of various materials*

Material	tonnes/m^3
Cement in bags	1.28
Cement in bulk	1.28 to 1.44
Wet sand	1.60
Damp sand	1.28
Aggregate	1.60

Additional material

In our calculations for the costs of concrete we can anticipate shrinkage of between 20 and 33.33 per cent. Allowances for the costs of additional material are determined as follows.

Shrinkage of 20 per cent

A volume of 1 m^3 less 20 per cent gives 0.80 m^3. Therefore to obtain 1 m^3 we shall have to add 0.20 m^3. This represents an increase of

$$\frac{0.20}{0.80} \times 100 = 25 \text{ per cent}$$

Shrinkage of 28.6 per cent

A volume of 1 m^3 less 28.6 per cent gives 0.714 m^3. Therefore to obtain 1 m^3 we shall have to add 0.286 m^3. This represents an increase of

$$\frac{0.286}{0.714} \times 100 = 40 \text{ per cent}$$

Shrinkage of 33.33 per cent

A volume of 1 m^3 less 33.33 per cent give 0.667 m^3. Therefore to obtain 1 m^3 we shall have to add 0.333 m^3. This represents an increase of

$$\frac{0.333}{0.667} \times 100 = 50 \text{ per cent}$$

BS 5328 Table 1. Mass of dry aggregate to be used with 100 kg of cement

Grade of concrete	Nominal maximum size of aggregate (mm)	40		20		14		10	
	Workability	Medium	High	Medium	High	Medium	High	Medium	High
	Range for standard sample (mm)	50–100	80–170	25–75	65–135	5–55	50–100	0–45	15–65
	Range for sample taken in accordance with 9.2 (mm)	40–110	70–180	15–85	55–145	0–65	40–110	0–55	5–75
		kg	kg	kg	kg	kg	kg	kg	kg
C7.5P		1080	920	900	780	N/A	N/A	N/A	N/A
C10P		900	800	770	690	N/A	N/A	N/A	N/A
C15P	Total	790	690	680	580	N/A	N/A	N/A	N/A
C20P	aggregate	660	600	600	530	560	470	510	420
C25P		560	510	510	460	490	410	450	370
C30P		510	460	460	400	410	360	380	320

N/A not applicable

BS 5328   Table 2. Percentage by mass of fine aggregate total aggregate

Grade of concrete	Nominal maximum size of aggregate (mm)	40		20		14		10	
	Workability	Medium	High	Medium	High	Medium	High	Medium	High
C7.5P C10P C15P		30–45		35–50		N/A		N/A	
C20P C25P C30P	Grading zone 1	35	40	40	45	45	50	50	55
	2	30	35	35	40	40	45	45	45
	3	30	30	30	35	35	40	40	45
	4	25	25	25	30	30	35	35	40

N/A

Notes on the use of Tables 1 and 2

NOTE 1. The proportions given in the tables will normally provide concrete of the strength in N/mm^2 indicated by the grade except where poor control is allied with the use of poor materials.

NOTE 2. For grades C7.5P, C10P and C15P a range of fine-aggregate percentages is given; the lower percentage is applicable to finer materials such as zone 4 sand and the higher percentage to coarser materials such as zone 1 sand.

NOTE 3. For all grades, small adjustments in the percentage of fine aggregate may be required depending on the properties of the particular aggregates being used.

NOTE 4. For grades C20P, C25P and C30P, and where high workability is required, it is advisable to check that the percentage of fine aggregate stated will produce satisfactory concrete if the grading of the fine aggregate approaches the coarser limits of zone 1 or the finer limits of zone 4.

Costing for shrinkage

Where the lowest percentage for shrinkage is used, it is practice to divide the cost for the concrete mix by the combined volumes of the fine and coarse aggregates. Where the higher percentages are used, it is practice to divide the cost for the concrete mix by the combined volumes of the cement and the fine and coarse aggregates.

Example 6.1

Ordinary prescribed mix: grade C20P concrete, 40 mm aggregate, medium workability. m^3

From Table 1 of BS 5328 we can see that for every 100 kg of cement we require 660 kg total aggregate. We must therefore refer to Table 2 of BS 5328 to establish the percentage by mass of fine aggregate; for grading zone 2 this is 30 per cent. Therefore the mix proportions are as follows:

Cement	100 kg
Fine aggregate	198 kg
Coarse aggregate	462 kg

The volume of mix based on the use of 1 tonne of cement is calculated as follows:

Cement	1.00 t/1.28 t/m^3	0.78
Fine aggregate	1.98 t/1.50 t/m^3	1.32
Coarse aggregate	4.62 t/1.60 t/m^3	2.89
Total volume of mix		4.99 m^3

We develop the costs for a mix based on the use of 1 tonne of cement. We assume that the cement is delivered in 50 kg bags, which are kept in a dry store and will require to be transferred from the store to the mixer. Aggregates, on the other hand, are stored in bulk adjacent to the mixer.

Cement	1.00 t at £60.00/t	£60.00
Fine aggregate	1.98 t at £7.00/t	13.86
Coarse aggregate	4.62 t at £7.00/t	32.34
Transfer 1 t cement: labourer 1 hour/t at £5.30		5.30
Cost of mix		111.50
Allow shrinkage 40%		44.60
Materials cost of 4.99 m^3		156.10
Material cost per m^3		£31.29

Example 6.2

> Ordinary prescribed mix: grade C25P concrete, 20 mm aggregate, medium workability.   m^3

From Table 1 of BS 5328 we can see that for every 100 kg of cement we require 510 kg total aggregate. We must therefore refer to Table 2 of BS 5328 to establish the percentage by mass of fine aggregate; for grading zone 2 this is 35 per cent. Therefore the mix proportions are as follows:

Cement	100 kg
Fine aggregate	178.5 kg
Coarse aggregate	331.5 kg

The volume of mix based on the use of 1 tonne of cement is calculated as follows:

Cement	1.00 t/1.28 t/m^3	0.78
Fine aggregate	1.78 t/1.50 t/m^3	1.19
Coarse aggregate	3.31 t/1.60 t/m^3	2.07
Total volume of mix		4.04 m^3

We develop the costs for a mix based on the use of 1 tonne of cement:

Cement	1.00 t at £60.00/t	£60.00
Fine aggregate	1.78 t at £7.00/t	12.46
Coarse aggregate	3.32 t at £7.00/t	23.24
Transfer 1 t cement: labourer 1 hour/t at £5.30		5.30
Cost of mix		101.00
Allow shrinkage 40%		40.40
Materials cost of 4.04 m^3		141.40
Material cost per m^3		£35.00

Mixer cost

Only small quantities of concrete will be mixed on site by hand. Larger quantities will arrive on site ready mixed by an independent producer, or will be mixed on site using a machine. Purchasing concrete from a producer eliminates the requirement to calculate the costs of material and mixing; only the costs for final transporting, placing etc. have to be added to the producer's quoted price.

Where a mixer is used, we must establish its cost. A typical method of assessing the mixer cost per hour is given in Example 3.1. In this chapter, mixer costs are given without calculation.

Labour costs

Having established how to calculate the costs for concrete mixes and mixers, we have now to determine the cost of labour required to produce, transport and place the concrete mix. As the essence of good concrete work is that continuous production be maintained, the number of men in a gang or squad is determined by the type of job to be done and the mixer being used.

We could adopt labour constants, that is average outputs for each type of concrete component which would account for producing, transporting and placing a mix. However, a more accurate way to develop the costs is to determine the output of a mixer in m^3 per hour, and to estimate the gang size required to mix, transport and place. The gang cost per m^3 is then calculated by dividing the gang cost per hour by the mixer output in m^3 per hour.

The rate of discharge is determined by the demand on the mixer on site and by its capacity in terms of number of batches per hour.

Mixer outputs and gang sizes

Figure 6.1 shows mixer outputs and related gang sizes. The gangs given on the graph are for mixing only and would be made up as follows:

One man gang Mixer operator.
Two man gang Mixer operator, feeder operator.
Three man gang Mixer operator, feeder operator, labourer.

It has been assumed that mechanically fed mixers are used, i.e. those with a feed hopper and dragline aggregate feeder. Where a mixer is fed manually, another labourer should be added to the squad.

Under the Working Rules agreement WR3, extra payments are made for extra skill or responsibility. Operatives responsible for a mixer (one per mixer) and for a dragline feeder are entitled to such extra payments, which are given in the Working Rules.

Placing outputs

Placing outputs are dependent on the type of concrete member to be poured and the mixer output. The gang size will therefore depend on these two factors. Table 6.2 indicates required gang sizes for a selection of typical members.

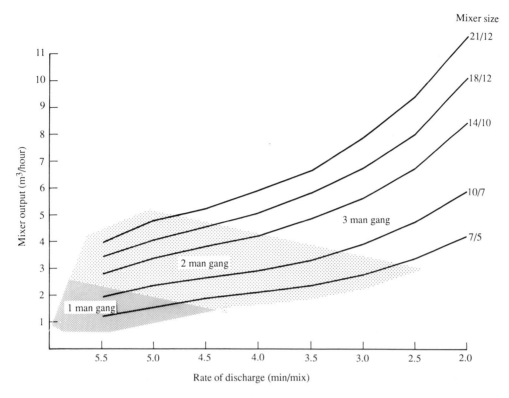

Figure 6.1 *Mixer outputs and related gang sizes*

Table 6.2 *Gang sizes required to pour particular members*

Member and description	Mixer output (m³/hour)		
	0–4	4–8	8–12
Foundations in trenches, over 300 mm thick	1	1	2
Foundations in trenches, not exceeding 300 mm thick	1	2	2
Beds not exceeding 150 mm thick	2	2	3
Beds 150–450 mm thick	1	2	2
Beds over 450 mm thick	1	1	2
Walls not exceeding 150 mm thick	2	3	3
Walls 150–450 mm thick	2	2	3
Walls over 450 mm thick	2	2	3
Columns	2	2	3
Staircases	2	3	3

Transporting outputs

Concrete is normally transported from its point of mixing to the required position on site by ordinary wheelbarrow or diesel powered dumper. On the average building site, where the terrain is rough and messy, a wheelbarrow is uneconomical for transporting concrete beyond about 60 m. A dumper can be used on distances of up to 400 m. However, the aggregates in a concrete mix can be settled out by the action of a dumper, and therefore it is better to minimize journey lengths.

A labourer transporting concrete using a wheelbarrow will do so at about 1 m³/hour over a 100 m round trip.

A standard dumper of 368 litres working at 20 per cent below maximum capacity can take full discharge from mixers up to 14/10 capacity. For larger mixers, a swivel skip dumper of say 510 litres would need to be used. Dumper costs can be calculated using the method adopted earlier for mixer costs. An average speed of 30 km/hour can be used in the calculation of dumper transport costs.

Other plant may be needed, such as a hoist to take a concrete mix from ground level to floor levels above. If this is the case then costs for such plant would have to be assessed and included in the rate build-up.

Example 6.3

> Plain *in situ* concrete in foundations: grade C25P, 20 mm aggregate, medium workability; poured on or against earth or blinded hardcore; not exceeding 300 mm thick.   m³

We can begin by taking the materials cost of the mix direct from Example 6.2.

Materials

Cost per m³	£35.00
Waste 5%	1.75

Mixer
We will use a 10/7 concrete mixer, and assume that the cost for this mixer is £0.59 per hour. From the mixer output graph (Figure 6.1) we can see that a 10/7 mixer discharging at say 3 minutes per mix will produce approximately 4 m³/hour. Therefore

Mixer cost per m³: £0.59/4   0.15

Mixing squad
From Figure 6.1 we see that the gang size required is 3 men. Therefore

Mixer operator cost/hour	£5.40
Feeder operator cost/hour	5.43
Labourer cost/hour	5.30
Gang cost/hour	16.13

Gang cost per m³: £16.13/4   4.03

Transporting
Transporting mix by wheelbarrow over say a 50 m round trip, assuming that 1 labourer will transport 1 m³/hour over 100 m:

Labourers (nr) to barrow 4 m³/hour: $(50/100) \times 4 = 2$
Cost to transport 4 m³: $2 \times £5.30 = £10.60$
Cost to transport 1 m³: £10.60/4   2.65

Placing
From Table 6.2 we see that 1 labourer is required to spread the mix. Therefore

Cost per m³ for 1 labourer at £5.30: £5.30/4	1.32
	44.90
Profit and oncost 20%	8.98
Rate per m³	£53.88

Example 6.4

> Plain *in situ* concrete beds: grade C25P, 20 mm aggregate, medium workability; poured on or against earth or blinded hardcore; 150–450 mm thick.   m³

Again we can take materials the cost of the mix from Example 6.2.

Materials

Cost per m³	£35.00
Waste 5%	1.75

Mixer
We will use a 14/10 concrete mixer, and assume that the cost for this mixer is £1.66 per hour. From the mixer output graph (Figure 6.1) we can see that a 14/10 mixer discharging at say 3 minutes per mix will produce approximately 5.75 m³/hour. Therefore

Mixer cost per m³: £1.66/5.75	0.29

Mixing squad
From Figure 6.1 we see that the gang size required is 3 men. Therefore

Mixer operator cost/hour	£5.40
Feeder operator cost/hour	5.43
Labourer cost/hour	5.30
Gang cost/hour	16.13
Gang cost per m³: £16.13/4	4.03

Transporting mix by wheelbarrow over say a 100 m round trip, assuming that 1 labourer will transport 1 m³/hour over 100 m:

Labourers (nr) to barrow 5.75 m³/hour: (100/100) × 5.75 = 5.75 (say 6)
Cost to transport 5.75 m³: 6 × £5.30 = £31.80

Cost to transport 1 m³: £31.80/5.75	5.53

Placing
From Table 6.2 we see that 2 labourers are required to spread the mix. Therefore

Cost per m³ for 2 labourers at £5.30: 2 × £5.30/5.75	1.84
	48.44
Profit and oncost 20%	9.69
Rate per m³	£58.13

Example 6.5

Reinforced concrete walls: grade C25P, 20 mm aggregate, medium workability; not exceeding 150 mm thick. Calculate the rate per m³.

The following costs per m³ are taken from Example 6.4:

Materials	£35.00
Waste 5%	1.75
Mixer	0.29
Mixing squad	4.03
Transporting	5.53

Placing costs are now added. From Table 6.2 we see that 3 labourers are required to spread the mix. Therefore

Cost per m³ for 3 labourers at £5.30: $3 \times £5.30/5.75$	2.77
	49.37
Profit and oncost 20%	9.87
Rate per m³	£59.24

Reinforcement for *in situ* concrete

Bar reinforcement

Reinforcement bars are generally delivered to site cut and bent in accordance with reinforcement drawings and bending schedules. Bars are fixed and secured in position using spacers and tie wire. An allowance for spacers is made in the examples that follow. Average quantities for tie wire per 1000 kg of reinforcement bar are given in Table 6.3.

Where cutting, bending and fixing bars is carried out on site, 5 per cent wastage should be added. On the other hand, where reinforcement is delivered to site cut and bent, 1 per cent should be added to allow for damage and loss.

Qualified bar benders and fixers are paid the craft operatives' wage in accordance with the Working Rules.

Table 6.4 gives labour constants for cutting and bending. Table 6.5 gives labour constants for fixing.

Table 6.3   *Average quantities of tie wire for reinforcement bar*

Reinforcement bar diameters (mm)	Tie wire per 1000 kg of bar, inc. waste (kg)
6–12	12
16–25	7
32–50	5

Table 6.4 *Labour constants for cutting and bending bar reinforcement (steel fixer hours per 1000 kg of steel bar)*

Concrete member	Reinforcement bar diameter (mm)									
	6	8	10	12	16	20	25	32	40	50
Foundations and beds	50	42	35	30	28	25	23	—	—	—
Slabs	50	42	35	30	28	25	23	—	—	—
Walls	50	42	35	30	28	25	23	—	—	—
Beams, columns and staircases	55	46	39	33	30	27	25	25	27	27
Beam and column casings	50	42	35	30	28	—	—	—	—	—
Links and stirrups	80	70	60	55	50	45	—	—	—	—

Table 6.5 *Labour constants for fixing bar reinforcement (steel fixer hours per 1000 kg of steel bar)*

Concrete member	Reinforcement bar diameter (mm)									
	6	8	10	12	16	20	25	32	40	50
Foundations and beds	50	42	35	30	28	25	23	—	—	—
Slabs	52	44	37	32	30	28	27	—	—	—
Walls	70	60	50	45	40	35	30	—	—	—
Beams, columns and staircases	70	60	50	45	40	35	30	28	26	24
Beam and column casings	60	58	42	36	33	—	—	—	—	—
Links and stirrups	90	80	70	55	50	45	—	—	—	—

Example 6.6

12 mm diameter hot rolled plain round mild steel bars in foundations, straight or bent.	tonne

Materials

1000 kg of 12 mm bars delivered cut and bent	£425.00
Waste 1%	4.25
	429.25
Unload and stack: labourer 1 hour at £5.30	5.30
	434.55
Tie wire: 12 kg at £14/kg	168.00
Chairs and spacers: say	10.00
	612.55

Labourer

Fixing: qualified steel fixer 30 hours at £5.90	177.00
	789.55
Profit and oncost 20%	157.91
Rate per tonne	£947.46

Example 6.7

20 mm diameter hot rolled deformed high yield steel bars in isolated beams, straight or bent.	tonne

Materials

1000 kg of 25 mm bars delivered in stock lengths	£335.00
Waste 1%	16.75
	351.75
Unload and stack: labourer 1 hour at £5.30	5.30
	357.05
Tie wire: 7 kg at £14/kg	98.00
Chairs and spacers, say	15.00
	470.05

Labour

Cutting, bending and fixing: qualified steel fixer 55 hours at £5.90	324.50
	794.55
Profit and oncost 20%	158.91
Rate per tonne	£953.46

Fabric reinforcement

Table 6.6 gives labour constants for unloading, stacking, cutting and laying fabric reinforcement.

Table 6.6   *Labour constants for fixing fabric reinforcement (labourer hours per unit)*

Concrete member or operation	Fabric reinforcement (kg/m²)										
	0.77	1.54	2.61	3.02	3.73	4.34	4.53	5.55	6.16	6.72	8.14
Unloading, stacking, square cutting, laying (per m²)											
Foundations	—	—	0.07	0.08	0.09	0.10	0.10	0.11	0.12	—	—
Beds and slabs	—	—	0.06	0.06	0.07	0.09	0.10	0.11	0.12	0.13	0.14
Walls	—	—	0.12	0.14	0.16	0.19	0.22	0.26	0.30	0.34	0.40
Staircases	—	—	0.16	0.20	0.25	0.31	0.37	0.44	0.50	0.56	0.65
Beam wrapping	0.15	0.19	—	—	—	—	—	—	—	—	—
Additional cutting (per metre)											
Raking cutting	0.05	0.06	0.08	0.09	0.11	0.13	0.14	0.15	0.16	0.17	0.18
Curved cutting	0.08	0.08	0.12	0.14	0.17	0.20	0.21	0.23	0.24	0.25	0.27

Example 6.8

Reinforcement: fabric; hard plain round steel, welded; reference A193; 200 mm side laps; 200 mm end laps.	m²

Materials

A193 fabric reinforcement: 1 m² at 3.02 kg/m² at £381/t	£1.15
Allowance for laps: add 12.5%	0.14
Wastage: add 5%	0.06
	1.35
Tie wire: allow 0.01 kg at £14.00/kg	0.14
Chairs and spacers: say	0.40
	1.89

Labour

Cutting and laying: labourer 0.08 hours at £5.30	0.42
	2.31
Profit and oncost 20%	0.46
Rate	£2.77

Example 6.9

Raking cutting on A193 fabric reinforcement.	m

Materials

Fabric reinforcement wasted by raking cutting: 0.25 m² at £1.15	£0.29

Labour

Cutting: labourer 0.09 hours/m at £5.30	0.48
	0.77
Profit and oncost 20%	0.15
Rate per m	£0.92

Formwork

The materials employed to construct formwork for *in situ* concrete are generally softwood boards, plywood and sheet steel for working faces, supported on a softwood or steel framework, often in conjunction with proprietary props. Timber is still widely used in the making of formwork because of the variety of forms that it will allow. Plywood has generally superseded timber boards for forming the faces of members.

Sheets or boards must be of sufficient thickness to take the weight of wet concrete. For plywood the thickness ranges from 25 to 50 mm depending on the structures to be formed and the spans involved.

Figure 6.2   *Formwork for columns, floors and beams*

Timber forms for *in situ* work are usually unfit for further use after four to six uses. In the production of precast units in a factory, timber formwork will have up to twenty uses.

In accordance with the Working Rules, a carpenter required to reuse materials for concrete work is entitled to an extra payment.

Examples 6.10–6.12 refer to the formwork structure shown in Figure 6.2.

Example 6.10

Formwork to soffits of horizontal slabs not exceeding 200 mm thick;
height to soffit exceeding 3 m but not exceeding 4.5 m.   m²

Area of soffit formwork between beams
= (4000 − 2 × 200) × (5000 − 2 × 200) = 16.56 m²

Materials

Plywood 19 mm: 16.56 m² at £7.27	£120.39
Softwood:	
175 × 50 mm joists: 22 m at £1.81	39.82
100 × 25 mm bearers: 9 m at £0.60	5.40

The props are taken in the beams only (Example 6.11), but their costs could be proportioned between floors and beams if felt necessary.

Nails: 0.5 kg at £3.50	1.75
	167.36
Wastage: add 10%	16.74
	184.10

Estimate 6 uses: cost per use £184.10/6	30.68
Shutter oil to formwork: say	1.50
Repairs and cleaning formwork: say	2.00
Cost per 16.56 m²	34.18

Material cost per m²: £34.18/16.56	2.06

Labour

Making, fixing and stripping: carpenter 2.7 hours at £6.15	16.61
	18.67
Profit and oncost 20%	3.73
Rate per m²	£22.40

Example 6.11

> Formwork to beams: attached to slabs; 400 × 400 mm; height to soffit exceeding 3 m but not exceeding 4.5 m.   m²

Area of beam formwork = 3 × 4600 × 400 = 5.52 m²

Materials

Plywood 19 mm: 5.52 m² at £7.27		£40.13
Softwood:		
225 × 47 mm sole plate:	9 m at £2.70	24.30
100 × 25 mm bracing:	6 m at £0.60	3.60
100 × 100 mm props:	7 m at £2.95	20.65
100 × 100 mm headtree:	3 m at £2.95	8.85
100 × 38 mm struts and battens:	5 m at £0.98	4.90
Folding wedges: 1 m at £0.98		0.98
Nails: 1 kg at £3.50		3.50
		106.91
Wastage: add 10%		10.69
		117.60

Estimate 5 uses: cost per use £117.60/5	23.52
Shutter oil to formwork: say	0.50
Repairs and cleaning formwork: say	1.50
Cost per 5.52 m^2	25.52
Material cost per m^2 £25.52/5.52	4.62

Labour

Making, fixing and stripping: carpenter 3.9 hours at £6.15	23.99
	28.61
Profit and oncost 20%	5.72
Rate per m^2	£34.33

Example 6.12

> Formwork to isolated columns: 400 × 400 mm; height to soffit exceeding 3 m but not exceeding 4.5 m.   m^2

Area of column formwork = 4 × 3500 × 400 = 5.6 m^2

Materials

Plywood 19 mm: 5.6 m^2 at £7.27	£40.71
Softwood:	
100 × 25 mm cleats: 7 m at £0.60	4.20
50 × 25 mm bearers: 2 m at £0.98	1.96
Nails: 1 kg at £3.50	3.50
	50.37
Wastage: add 10%	5.04
	55.41
Estimate 8 uses: cost per use £55.41/8	6.93
Steel clamps: 7 sets at £8.50/set	
Estimate 40 uses: cost per use (£8.50 × 7)/40	1.49
Repairs and cleaning formwork: say	1.80
Cost per 5.6 m^2	10.22
Material cost per m^2 £10.22/5.6	1.83

Labour

Making, fixing and stripping: carpenter 4.2 hours at £6.15	25.83
	27.66
Profit and oncost 20%	5.53
Rate per m^2	£33.19

Precast concrete

As stated in SMM7 in the coverage rules for precast concrete members, unit rates are deemed to include moulds, reinforcement, bedding, fixings, and so on. The unit rates developed in the examples that follow are for regular sections; however, the methods can be adapted to develop costs for irregular or one-off units.

The costs are for members produced on site by the contractor; therefore the rate build-ups used previously are adopted again.

Example 6.13

Before calculating the rates for individual precast concrete lintels, it is normal practice to calculate the cost of 1 m³ concrete for the various types of lintel itemized in the bill: for example, plain rectangular lintels, all faces left rough, reinforced with two 10 mm hot rolled plain mild steel bars. One hundred lintels, 100 mm × 100 mm × 1 m, can be cast from 1 m³ concrete.

Concrete
The following costs per m³ are taken from Example 6.4:

Materials	£35.00
Waste 5%	1.75
Mixer	0.29
Mixing squad	4.03
Labour placing concrete in mould: 10 hours at £5.30	53.00
Steel bars 10 mm: 200 m at 0.616 kg/m at £0.43/kg	52.98
	147.05

Mould

Plywood 19 mm: 0.32 m² at £7.27	£2.33	
Mould clamps: 3 nr at £0.20	0.60	
Mould oil etc.: say	0.10	
Carpenter making mould: 7 hours at £6.10	42.70	
	45.73	
Allow 20 uses per mould: cost per use £45.73/20 = £2.29		
At 1 use per lintel for 100 lintels: 100 × £2.29		228.63
Cost to cast 100 nr lintels 1 m long (1 m³)		£375.68

Having established the cost of 1 m³ of material for plain rectangular section lintels, we can calculate the cost for lintels of a similar pattern but with different size of section by using the ratio of area of cross-section to 1 m². This is illustrated in the next example.

Example 6.14

> Lintels: plain rectangular shape; reinforced concrete; 100 × 150 × 1000 mm long; reinforced with 1.23 kg, 10 mm bars. Nr

Materials

Cost of 1 m³ lintels 1 m long is £375.68 (Example 6.13)	
Cost of a 100 × 150 mm lintel 1 m long is £375.68 × (100 × 150)/ 1000 × 1000)	£5.64
Mortar bed for lintel: say	0.30
Wastage 2.5%	0.15

Labour

Labourer stacking on site: 0.08 hours at £5.30	0.42
Labourer transferring lintel from stack and hoisting to position on scaffold: 0.10 hours at £5.30	0.53
Bricklayer and labourer lifting into position, levelling and bedding: 0.16 hours at £11.40	1.82
	8.86
Profit and oncost 20%	1.77
Cost per m	10.63
Rate per lintel 1 m long: £10.61 × (1000/1000)	£10.63

Example 6.15

> Sills: sunk weathered; grooved; finished smooth 190 mm girth; 175 × 75 × 600 mm long; reinforced 0.37 kg 10 mm bars.   Nr

In this example we shall presume that the sills have been purchased from a manufacturer and delivered to the site.

Materials

Cost of 175 × 75 × 600 mm sill delivered to site	£5.80
Mortar bed for sill: say	0.20
Wastage 2.5%	0.15

Labour

Labourer unloading and stacking on site: 0.07 hours at £5.30	0.37
Labourer transferring sill from stack and hoisting to position on scaffold: 0.10 hours at £5.30	0.53
Bricklayer and labourer lifting into position, levelling and bedding: 0.13 hours at £11.40	1.48
	8.53
Profit and oncost 20%	1.71
Rate per sill	£10.24

7

Brickwork and blockwork

SMM7 Sections F10, F30

Introduction

Bricks can be obtained in a number of sizes and various compositions, strengths and shapes, to mention only the more obvious factors. The appropriate British Standards give details for clay, concrete, sand-lime and other bricks.

To give examples of rates using all of the different types of bricks would be impossible in the space allocated, as well as highly repetitive. The examples in this chapter will concentrate on a metric brick with dimensions 215 mm long, 102.5 mm wide and 65 mm thick. Whether the brick is solid, hollow or perforated, or has single or double frogs, will make no difference to us here, although in theory the different weights of bricks could affect the laying output per man and other factors.

Mortar

If bricks have no perforations or frogs, are absolutely rectilinear, are laid on a bed of mortar 10 mm thick and have each end jointed 10 mm thick, then 1000 bricks require the following volume of mortar:

$$1000[(0.215 \times 0.1025 \times 0.01) + (0.1025 \times 0.065 \times 0.01)] = 0.287 \, \text{m}^3$$

The amount of mortar taken by estimators varies from 0.5 to 0.8 m³. This makes allowance for the mortar which disappears into the frogs, hollows or perforations in the bricks, and also for the large amount of waste generated. This waste occurs throughout the process: cement, lime and sand are dumped on the ground and trodden in; mortar is left in the drum of the mixer, in the barrow, on the mortar board and on the scaffold; and mortar is squeezed out of the bed and joint of every brick. The allowance is large, but there is a lot of waste!

The mixes and costs of mortars used in the examples in this chapter are given in the appendix to the book.

Bricks per wall area

Obviously we need to know the number of bricks to be laid in each square metre of wall. For a half brick wall laid using metric bricks with dimensions $215 \times 102.5 \times 65$ mm and mortar joints of 10 mm, the number of bricks required per square metre is given by

$$\frac{1000 \times 1000}{225 \times 75} = 59.25925 \simeq 59$$

One brick walls need 118 bricks; one and a half brick walls need 177 bricks; and so on.

The waste on bricks can vary as with any other material, but in this chapter we will allow a modest 5 per cent.

Labour costs

Labour costs depend first on the squad make-up, i.e. the numbers of craftsmen and labourers. In the examples in this chapter, the squad comprises two bricklayers and one labourer.

The second factor is the number of bricks which each bricklayer can lay per hour. Fanciful claims are common on building sites; tales abound of bonuses earned, records shattered and employers dumbfounded. The truth is much less in magnitude. A good average – every day, 5 days a week, 47 weeks per annum – is 55–65 common bricks per hour. Obviously this is in straight runs of walling without excessive cutting for bond and without anything unusual. Output might be a little higher on thicker walls simply because the bricklayers and their labourer are spending proportionately less time moving along the wall.

Common brickwork

Example 7.1

Common brickwork half brick thick in cement mortar 1:3, built in stretcher bond.	m^2

Common bricks $215 \times 102.5 \times 65$ mm delivered to site: per 1000		£105.00
Wastage 5%		5.25
		110.25
Labour: 2 bricklayers per hour at £6.10	£12.20	
1 labourer per hour at £5.30	5.30	
	17.50	
Output per tradesman 55 bricks/hour; 2 tradesmen:		
Cost labour per 1000 bricks: £17.50 × 1000/(2 × 55)		159.09
Mortar at £62.79/m³ at 0.53 m³ per 1000 bricks for half brick wall		33.28
Cost for 1000 bricks		302.62
Number bricks per m² for half brick thick wall is 59		
Cost per m²: £302.62 × (59/1000)		17.85
Profit and oncost 20%		3.57
Rate per m²		£21.42

Example 7.2

Common brickwork one brick thick in cement mortar 1:3, built in English bond.	m^2

Common bricks $215 \times 102.5 \times 65$ mm delivered to site: per 1000		£105.00
Wastage 5%		5.25
		110.25

Labour: 2 bricklayers per hour at £6.10	£12.20	
1 labourer per hour at £5.30	5.30	
	17.50	

Output per tradesman 55 bricks/hour; 2 tradesmen:
Cost labour per 1000 bricks: £17.50 × 1000/(2 × 55)   159.09

Mortar at £62.79/m³ at 0.60 m³ per 1000 bricks for one brick wall   37.67

Cost for 1000 bricks   307.01

Number bricks per m² for one brick thick wall is 118
Cost per m²: £307.01 × (118/1000)   36.23
Profit and oncost 20%   7.25

Rate per m²   £43.48

Example 7.3

Common brickwork one and one half brick thick in cement/lime mortar 1:1:6, built in Flemish bond.	m²

Common bricks 215 × 102.5 × 65 mm delivered to site: per 1000	£105.00
Wastage 5%	5.25
	110.25

Labour: 2 bricklayers per hour at £6.10	£12.20	
1 labourer per hour at £5.30	5.30	
	17.50	

Output per tradesman 60 bricks/hour; 2 tradesmen:
Cost labour per 1000 bricks: £17.50 × 1000/(2 × 60)   145.83

Mortar at £50.58/m³ at 0.60 m³ per 1000 bricks 1.5 brick wall   30.35

Cost for 1000 bricks   286.43

Number bricks per m² for 1.5 brick thick wall is 177
Cost per m²: £286.43 × (177/1000)   50.70
Profit and oncost 20%   10.14

Rate per m²   £60.84

Brickwork sundries

Example 7.4

Form cavity 75 mm wide including 4 butterfly wall ties per m².	m²

Cost 4 wall ties per m² at £37 per 1000	£0.15
Wastage 5%	0.01
Tradesman sets out cavity at 12 m²/hour at £6.10/hour:per m²	0.51
	0.67
Profit and oncost 20%	0.13
Rate per m²	£0.80

Example 7.5

> Hessian based bituminous sheet in horizontal damp-proof course
> lapped at joints and corners, in width not exceeding 225 mm.   m

Most authorities apply a standard cost for laying this material irrespective of width, although SMM7 does require the quantity surveyor to keep the classifications separate. The labour output for vertical DPC is half that for horizontal DPC.

DPC costs £3.80 per m²	£3.80
Waste and laps 5%	0.19
Tradesman at £6.10 per hour lays 5 m²:per m²	1.22
Labourer at £5.30 per hour lays 5 m²:per m²	1.06
	6.27
Profit and oncost 20%	1.25
Rate per m²	7.52
Rate per m at 225 mm wide	£1.69

Example 7.6

> Hessian based bituminous sheet complying with BS 743 type A, in
> a vertical damp-proof course lapped at joints and corners, in width not
> exceeding 225 mm.   m

DPC costs £3.80 per m²	£3.80
Waste and laps 5%	0.19
Tradesman at £6.10 per hour lays 2.5 m²:per m²	2.44
Labourer at £5.30 per hour lays 2.5 m²:per m²	2.12
	8.55
Profit and oncost 20%	1.71
Rate per m²	10.26
Rate per m at 225 mm wide	£2.31

Example 7.7

Galvanized expanded metal brickwork reinforcement.	m

Material per 25 m roll 100 mm wide (1 roll = 2.5 m²)	£8.00
Waste 5%	0.40
Tradesman at £6.10 per hour lays 2.5 m²	6.10
Labourer at £5.30 per hour lays 2.5 m²	5.30
	19.80
Profit and oncost 20%	3.96
Rate per roll	23.76
Rate per m at 100 mm wide	£0.95

Facing brickwork

Example 7.8

One and one half brick thick wall built with common bricks, faced one side with facing bricks PC £212 per 1000 delivered to site, built in Flemish bond and jointed in cement/lime mortar 1:1:6.	m²

Cost of facing bricks per 1000 delivered to site		£212.00
Unload and stack: labourer 2 hours at £5.30		10.60
		222.60
Wastage 5%		11.13
		233.73
Labour: 2 bricklayers per hour at £6.10	£12.20	
1 labourer per hour at £5.30	5.30	
	17.50	
Output per tradesman 40 bricks/hour: 2 tradesmen		
Cost labour per 1000 bricks: £17.50 × 1000/(2 × 40)		218.75
Mortar at £50.58/m³ at 0.60 m³ per 1000 bricks for 1.5 brick thick wall		30.35
Total cost for building 1000 facing bricks		482.83
Number of facing bricks per m² in Flemish bond is 79		
Cost of facing brick per m²: £482.83 × (79/1000)		38.14
Bricklayer joints 2.4 m²/hour at £6.10: per m²		2.54
Number of bricks per m² for 1.5 brick wall is 177: so commons 177−79=98		
Commons in same mortar cost £286.43 per 1000 (Example 7.3): 98 cost		28.07
		68.75
Profit and oncost 20%		13.75
Rate per m²		£82.50

Example 7.9

> Plain brick-on-edge horizontal cope one and one half brick wide,
> formed with facing bricks PC £212 per 1000 delivered to site, built and
> jointed in cement/lime mortar 1-1-6.   m

Cost of facing bricks per 1000 delivered to site	£212.00
Unload and stack: labourer 2 hours at £5.30	10.60
	222.60
Wastage 5%	11.13
	233.73

Labour: 2 bricklayers per hour at £6.10	£12.20	
1 labourer per hour at £5.30	5.30	
	17.50	

Output per tradesman 30 bricks/hour: 2 tradesmen:

Cost labour per 1000 bricks: £17.50 × 1000/(2 × 30)	291.67
Mortar at £50.58/m³ at 0.60 m³ per 1000 bricks for 1.5 brick wall	30.35
Total cost of building 1000 facing bricks	555.75

Coping with 65 mm thick bricks:

65 mm thick brick + 10 mm joint = 13.33 bricks/m at 1 brick thick, say 21 bricks/m   at 1.5 bricks
thick

Cost £555.75 × (21/1000)	11.67

Bricklayer joints 2.4 m²/hour at £6.10
Girth of cope 112.5 + 327.5 + 112.5 = 553 mm

Area jointed per metre is 0.55 m²: cost £6.10 × (0.55/2.4)	1.40
	13.07
Profit and oncost 20%	2.61
Rate per metre	£15.68

Jointing and pointing

In Examples 7.8 and 7.9 the brickwork was described as *jointed*. This is the term applied to the finishing of the joints between bricks, usually with a struck weathered joint using the same mortar as for building. The most important feature of jointing is that it is done as the bricks are built, the mortar in each bed or joint being formed while still plastic into the struck weathered finish. No material is added at a later date.

Pointing is the term used to describe the procedure where a separate mortar – very often of a different mix and even colour from the building mortar – is applied to the joints between the bricks. As the bricklayer lays the bricks he will rake out a little of the bedding mortar to leave space for this additional application. The finish achieved at the joints may be struck weathered but there are many other finishes, and the reader is referred to the many books on brickwork currently available.

In Example 7.8 we could add pointing with a white mortar made from Snowcrete cement 1:4. We have no calculation for this mortar but will price arbitrarily at £120 per m³. The calculation for the new rate would therefore be as follows.

Example 7.10

One and one half brick thick wall built with common bricks, faced one side with facing bricks at £212 per 1000 delivered to site, built in Flemish bond in cement/lime mortar 1:1:6 and pointed in white mortar made from Snowcrete and silver sand 1:4.	m²

Cost of bricks as Example 7.8	£233.73
Cost of labour as Example 7.8	218.75
Mortar to build as Example 7.8	30.35
Mortar to point 1000 bricks: 0.033 m³ at £120 per m³	3.96
Total cost of building 1000 facing bricks	486.79
Number of facing bricks per m² in Flemish bond is 79	
Cost of facing brick per m²: £486.79 × (79/1000)	38.46
Bricklayer rakes and points joints 1 m² hour £6.10	6.10
	44.56
Cost 98 commons as Example 7.8	28.07
	72.63
Profit and oncost 20%	14.53
Rate per m²	£87.16

Special brickwork

Example 7.11

Facing brickwork in horizontal projection 56 mm deep from face of wall and 440 mm high, built brick-on-end in cement/lime mortar 1:1:6.	m

Cost of facing bricks per 1000 including waste as Example 7.8	£233.73
Squad cost £17.50 per hour as Example 7.8 Cost labour per 1000 bricks: £17.50 × 1000/(2 × 25)	350.00
Mortar at £50.58/m³ at 0.60 m³ per 1000 bricks for 1.5 brick wall	30.35
Total cost of building 1000 facing bricks	614.08
Projection with 65 mm thick bricks has 26.66 bricks/m, say 27 Cost £614.08 × (27/1000)	16.58

To produce a projection of 56 mm requires that a cut brick be built into the wall behind the projecting facing brick. The cut bricks are of course common bricks. It is assumed that one common brick will not be cut into two usable pieces; this covers waste in the cut brickwork.

Commons in same mortar cost £286.43 per 1000 (Example 7.3): 27 cost	7.73

Additional labour rough cutting common brickwork 0.44 m² per m of projection

Bricklayer cuts 1 m²/hour at £6.10/hour

Bricklayer cuts 1 m²/hour at £6.10/hour	2.68

Bricklayer joints 2.4 m²/hour at £6.10
Girth of projection $440 + (2 \times 56) = 552$ mm

Area jointed per metre is 0.55 m²: cost £6.10 × (0.55/2.4)	1.40
	28.39
Profit and oncost 20%	5.68
Rate per metre	£34.07

Example 7.12

> Facing brick semi-circular arch, one brick wide on face and one and one half bricks wide on soffit, with purpose made facing brick voussoirs, delivered to site at £450 per 1000, built in cement/lime mortar 1:1:6 and jointed.   m

Cost of facing bricks per 1000 delivered to site	£450.00
Unload and stack: labourer 2 hours at £5.30	10.60
	460.60
Wastage 5%	23.03
	483.63

Squad cost £17.50 per hour as Example 7.8
Output per tradesman 20 bricks/hour: 2 tradesmen

Cost labour per 1000 bricks: £17.50 × 1000/(2 × 20)	437.50
Mortar at £50.58/m³ at 0.60 m³ per 1000 bricks for 1.5 brick wall	30.35
Cost of laying 1000 voussoirs	951.48

Mean girth or length on face: assume that the bricks are 75 mm at the widest part (the extrados) and 55 mm wide at the narrowest part (the intrados). The mean width of each voussoir is then 65 mm. Allowing a 10 mm joint, the number of bricks per metre of arch is 13.33 for a half brick thick wall. Say 40 bricks per metre are required for a 1.5 brick arch. Then cost per metre £951.48 × (40/1000)

	38.06

Bricklayer joints 1.5 m² per hour at £6.10
Girth of arch $327.5 + (2 \times 225) = 777.5$ mm (2 × faces and soffit)

Area jointed per metre is 0.78 m²: cost £6.10 × (0.78/1.5)	3.17
	41.23
Profit and oncost 20%	8.25
Rate per metre	£49.48

Blockwork

Blocks are manufactured in a variety of materials, clay and concrete being the most usual. They can be hollow, joggle jointed, and filled with foamed plastic, and may be finished on one or both faces with a variety of decorative finishes or left plain to receive plaster. Like bricks they can be built fair faced, and are very often painted direct when built this way. Building blocks with a fair face or a pre-applied decorative finish will lower output and will also introduce the need for jointing or pointing. The latter are treated in the same way as for brickwork; the bricklayer can point a larger area of blockwork per hour than of brickwork.

Example 7.13

75 mm thick lightweight aggregate concrete block complying with BS 6073: Part 1, built in cement/lime mortar 1:1:6.	m²

Material per m² delivered to site and mechanical off-loaded (MOL)	£4.80
Waste 5%	0.24
Squad cost £17.50 per hour as Example 7.8 Output per bricklayer 1.5 m²/hour: 2 bricklayers Cost labour per m²: £17.50/(2 × 1.5)	5.83
Mortar at £50.58/m³ at 0.007 m³ per m² of 75 mm blockwork	0.35
	11.22
Profit and oncost 20%	2.24
Rate per m²	£13.46

Example 7.14

100 mm thick autoclaved aerated concrete block complying with BS 6073: Part 1, built in cement/lime mortar 1:2:9.	m²

Material per m² delivered to site and (MOL)	£6.10
Waste 5%	0.30
Squad cost £17.50 per hour as Example 7.8 Output per bricklayer 1.5 m²/hour: 2 bricklayers Cost labour per m²: £17.50/(2 × 1.5)	5.83
Mortar at £50.64/m³ at 0.01 m³ per m² of 100 mm blockwork	0.51
	12.74
Profit and oncost 20%	2.55
Rate per m²	£15.29

8

Underpinning
SMM7 Sections C30, D50

Introduction

Underpinning is the transference of the load of a building to a lower stratum of the ground. This is done for several reasons:

- To permit the construction of a new basement within or adjacent to the existing foundations which are above the level of the new basement
- To provide a new foundation at a lower level than the existing, where the existing has failed due to, for example, shrinking clay or subsidence
- To increase the load bearing capacity of the existing foundation
- In extreme cases, to allow the whole building to be mechanically moved by installing a ground beam.

There are a number of ways to carry out the underpinning operation; the choice depends upon the stability of the existing structure. Except in the most simple cases and where the structure is in good condition, the work will be carried out by a specialist contractor. This contractor will employ stabilizing techniques such as the drilling of ground anchors and/or the construction of ground beams.

For a straightforward operation on a stable structure the contractor will price the work described in the bills of quantities. The constants used will be based upon those for the ground work, *in situ* concrete and masonry sections.

Shoring (C30)

It may be necessary for the estimator to price an item in the bills of quantities for strutting to window and door openings and also for temporary shoring to walls. The extent of this work will be described as an item and it will be necessary for the estimator to have this temporary structural work designed before pricing.

Underpinning (D50)

The excavation work in underpinning is in two parts: the preliminary trench from the existing surface level to the base of the foundation, and the underpinning pits below the existing foundation. The preliminary trench will be priced in an identical manner to that described for hand excavation in Chapter 5.

The underpinning pits are the excavations below the existing foundation level undertaken by hand on a section of foundation normally 1 m to 1.5 m long. The working on such small sections, in part below the existing foundation, results in an addition to the hand excavation constant of 100 per cent. The bringing of the soil to the surface in the number of lifts calculated by dividing the total depth to the base of the underpinning by 1.5 m, as in Chapter 5, is calculated at the given constants plus 50 per cent.

Where underpinning is carried out by working from one side only, the earthwork support to the back face of the excavation under the existing foundation is left in. This section of earthwork support is commonly in precast concrete planks, each with a hole in the top to allow for grouting any void left behind the earthwork support (see Example 8.2).

For all remaining items in the underpinning section, i.e. groundwork, *in situ* concrete and masonry, the constants used are those of the appropriate chapter plus 50 per cent.

Example 8.1

> Excavating underpinning pits, maximum depth not exceeding 2 m, from one side only. m^3

The difficulty multipliers are obtained from Table 5.2, and the outputs (man hours per m^3) from Tables 5.3 and 5.4.

	Volume excavated	Difficulty multiplier	Man hours	Rate per hour	Cost
To 1.5 m deep	1.5	3.00	2.2	£5.30	£52.47
To 2 m deep	0.5	3.00	2.2	£5.30	17.49
Throw one stage	0.5	2.00	1.4	£5.30	7.42
Total volume	2.0	Total cost			77.38

Divide total cost by total volume	38.69
Profit and oncost 20%	7.74
Rate per m^3	£46.43

Example 8.2

> Earthwork support to underpinning pits left in, maximum depth not exceeding 2 m, distance between opposing faces 2–4 m. m^2

Materials
Precast reinforced concrete plank 50 mm × 200 mm wide × 2 m high:

5 nr at £4.50	£22.50
Waste 5%	1.13

Walings 50 × 200 mm:		
3 nr × 1 × 0.05 × 0.20 m at £220/m^3	£6.60	
Strutting 100 × 100 mm (to ground) at 2 per metre = 3 pairs:		
6 nr × (av.) 1.2 × 0.10 × 0.10 m at £220/m^3	15.84	
Base 100 × 100 × 200 mm long (base of struts):		
6 nr × 0.20 × 0.10 × 0.10 m at £220/m^3	2.64	
	25.08	

Divided by 3 uses   8.36
Waste 10%   0.84

Labour
Unload, store and place in position precast concrete planks:
  5 nr at 12 minutes each at £5.30 per hour
                                                                     5.30

Unload, store, set up and strike timberwork:
  0.11 m^3 at 10 hours/m^3 at £5.30 per hour   5.83

                                                                    43.96
Profit and oncost 20%   8.79

Rate per m^2 £52.75

9

Roof coverings
SMM7 Sections H60, H62, H63, H71

Slate roofing

The main sources for roofing slates in Britain are Cumbria, Cornwall and North Lancashire in England, and Bangor, Portmadoc and Caernarvon in Wales. There are no longer large scale quarries in Scotland; therefore the main source of Scottish slates is from demolition for use second hand. On the British market there are also roofing slates from European countries such as Spain and Portugal. Roofing slates are sold by the tonne or by the thousand depending upon the source.

British Standard BS 680 Part 1:1944 (imperial) and Part 2:1971 (metric) details geological formations of true slate rock from which roofing slates can be quarried, together with characteristics, grade tests and sizes. The labour outputs given here are for standard size slates in accordance with BS 680 Part 2 Table 1, 'Standard lengths and widths of slates'. Sizes of randoms and peggies can be obtained from Table 2, 'Range of lengths for randoms and peggies'.

The labour outputs in Table 9.1 are based on a squad of two slaters and one labourer, and allow for carrying and fixing in position. Unloading, holing and dressing of slates are highlighted as required in the examples which follow.

The wastage on slates is taken as 2.5 per cent for sized and 5 per cent for unsized slates.

Under the rules laid down in SMM7, roof and wall coverings are measured in square metres, and items are deemed to include underlay and battens. These items have therefore to be accounted for under the superficial items for slating and tiling where required.

Tables 9.2 and 9.3 show expected slater outputs for a range of tasks.

In the following examples, in order to calculate the cost per m^2 we have first to establish the covering capacity of one slate. To do this we have to calculate the gauge of the slate and multiply by its width.

Table 9.1   *Range of slater outputs fixing slates*

Slate size (mm)	Output per tradesman (m²/hour)	Slate size (mm)	Output per tradesman (m²/hour)
610 × 355		355 × 205	
610 × 305	5.0–6.0	355 × 180	1.5–2.5
560 × 305		330 × 280	
560 × 280		330 × 255	
510 × 305	4.0–4.5	330 × 205	1.5–2.5
510 × 255		330 × 180	
460 × 305		305 × 255	
460 × 255	2.0–3.0	305 × 205	1.5–2.0
460 × 230		305 × 155	
405 × 305		255 × 255	
405 × 255	2.0–3.0	255 × 205	1.25–1.5
405 × 230		255 × 150	
405 × 205			
355 × 305	2.0–3.0		
355 × 255			

Table 9.2   *Range of slater outputs for labour items*

Measured items	Output per hour per tradesman
Square cutting around openings	3–4 m
Angle cutting to valleys	4–6 m
Angle cutting to hips	8–12 m
Double course at **eaves**	14–18 m
Square verges including bedding	6–10 m
Holes for pipes	3–4 nr
Ridge capping including bedding	5–6 m
Hip capping including bedding	4–5 m
Wastage on labour items:	
Square cutting	add 33.33%
Raking cutting	add 50%

Table 9.3 *Range of slater outputs for associated works*

Associated works	Output per hour per tradesman
Underslating felt	2.5–3.5 rolls of 15 m^2
Battens	35–45 m

Example 9.1

> Roofing: best quality second hand Scottish slates to BS 680; random sizes 355 to 255 mm long in diminishing courses: fixing with two 40 mm galvanized steel nails per slate to 75 mm lap on reinforced underslating felt to softwood sawn sarking, pressure impregnated with preservative; coverings to sloping pitch 40 degrees.   m^2

Material: slates

Second hand slate delivered to site at £205 per 1000	£205.00
Take delivery and stack: 1.5 hours per 1000 at £5.30	7.95
Nails: 40 mm galvanized at £3.50/kg at 3.18 kg per 1000: 2000 nails per 1000 slates	22.26
Labour preparing slates per 1000:	
Tradesmen reholing 4 hours, redressing 2 hours: 6 hours at £6.10	36.60
Labourer sizing: 2 hours at £5.30	10.60
	282.41
Allow waste 5%	14.12
Cost per 1000 slates prepared ready for fixing	296.53

Slate gauge = [average length − (lap + 25)]/2 mm
          = [305 − (75 + 25)]/2 = 102.5 mm = 0.1025 m
Coverage 1 slate = width × gauge = 0.205 × 0.1025 = 0.021 m^2
Coverage 1000 slates = 0.021 × 1000 = 21.01 m^2
Cost per m^2 = £296.53/21.01   14.11

Labour: fixing slates
From Table 9.1, for the average slate size of 305 × 205 mm, 1.75 m^2 can be carried and fixed in position per hour per tradesman. Assume a squad of 2 tradesmen and 1 labourer.
Labour cost per hour: (2 × £6.10) + (1 × £5.30) = £17.50
Cost per m^2 £17.50/(2 × 1.75)   5.00

Slating cost per m^2 19.11

In accordance with SMM7 coverage rules for coverings, underlay and battens are deemed to be included. Therefore, where a specification requires either of those items, their costs have to be incorporated in the rate for the coverings as follows.

Materials: *underslating felt*

Felt 15 m² roll delivered to site	£13.24
Unload and stack: 0.05 hours at £5.30	0.26
	13.50

Nails fixed at 300 mm centres: roll length 15 m + width 1 m = 16 m
Number of nails per roll: 16/0.3 = 53.33, say 54

Cost of 54 nails at £2.80/kg at 2.85 kg per 1000	0.43
	13.93
Allow for waste 5%	0.70
	14.63

Labour: *fixing underslating felt*
Squad of 2 tradesmen and 1 labourer fixes 6 rolls per hour
Labour cost per hour: £17.50

Cost to fix 1 roll: £17.50/6	2.92
Cost to lay 1 roll	17.55

To calculate the cost for 1 m² of felt we must first determine the effective coverage of 1 roll. That is, we must subtract the laps from the roll dimensions, as follows:
Coverage = $(15 - 0.15) \times (1 - 0.075) = 14.85 \times 0.925 = 13.74$ m²

Cost to lay 1 m² felt: £17.55/13.74	1.28
	20.39
Profit and oncost 20%	4.08
Rate per m²	£24.47

Example 9.2

> Roofing; natural slates, Penrhyn heather blue standard grade; size 405 × 255 mm; fixing with two copper nails per slate to 75 mm lap to softwood battens 38 × 25 mm pressure impregnated with preservative, on reinforced underslating felt; coverings sloping at 30 degree pitch.   m²

Materials: *slates*

Heather blue slate ex quarry at £610 per 1000	£610.00

Assume site 200 miles from quarry
700 slates per crate weighing 1.06 tonnes per crate: by 16 tonne lorry
Nr of crates in load: 16/1.06 = 15.09, say 15 crates
Nr of slates in load: 15 × 700 = 10,500

Loading lorry:	1		
Unloading at site:	1		
Travelling time 1 man:	10		
Rest time etc.:	8		

20 hours at £5.51	£110.20	
Lorry cost: 20 hours at £7.50	150.00	
Delivery cost of full load 10,500 slates	260.20	
Delivery cost for 1000 slates: £260.20/10.5		24.78
Take delivery and stack: 1.5 hours per 1000 at £5.30		7.95
Nails: 40 mm copper at £5.50/kg at 2.7 kg per 1000: 2000 nails per 1000 slates		29.70
Labour: preparing slates per 1000: holing 4 hours at £6.10		24.40
		696.83
Allow for waste 5%		34.84
Cost per 1000 slates prepared ready for fixing		731.67

Material: fixing slates
Slate gauge $= [\text{length} - (\text{lap} + 25)]/2 \text{ mm}$
$\qquad = [405 - (75 + 25)]/2 = 152.5 \text{ mm} = 0.1525 \text{ m}$
Coverage 1 slate $=$ width $\times$ gauge $= 0.255 \times 0.1525 = 0.03889 \text{ m}^2$
Coverage 1000 slates $= 0.03889 \times 1000 = 38.89 \text{ m}^2$
Cost per m^2: £731.67/38.89   18.81

Labour: fixing slates
From Table 9.1, for the slate size of 405×255 mm, 2.5 m^2 can be carried and fixed in position per hour per tradesman. Assume a squad of 2 tradesmen and 1 labourer.
Labour cost per hour: $(2 \times £6.10) + (1 \times £5.30) = £17.50$
Cost per m^2: £17.50/(2×2.5) 3.50

Slating cost per m^2 22.31

Underslating felt as Example 9.1: per m^2 1.28

Materials: battens
Nr of slates per m^2: $1000/38.89 = 25.72$
Quantity 38×25 mm battens per $\text{m}^2 =$ nr slates per $\text{m}^2 \times$ slate width
$\qquad\qquad = 25.72 \times 0.255 = 6.56 \text{ m}$
Cost of battens per m^2: 6.56 m at £0.28 per metre   £1.84

Nr nails per $\text{m}^2 = $ (length battens)/(rafter centres) $= 6.56/0.45$
Weight nails per m^2 at 260 nails/kg: $(6.56/0.45)/260 = 0.056$ kg
Cost of nails per m^2: 0.056 kg at £2.50/kg   0.14

	1.98	
Allow for waste 5%	0.10	2.08

Labour: battens
Squad of 2 tradesmen and 1 labourer fixes 80 m battens per hour
Labour cost per hour: £17.50
Cost to fix battens for 1 m^2 slating: £17.50 $\times$ (6.56/80)   1.43

	27.10
Profit and oncost 20%	5.42
Rate per m^2	£32.52

Example 9.3

> Eaves: double course of 405 × 255 mm heather blue slates at eaves.   m

Materials
Cost 1000 heather blue slates from Example 9.2   £731.67

As the slates arrive on site pre-sized at 405 × 255 mm, those at the eaves have to be
dressed to a length equal to gauge + lap + 25 mm
Dressing to size: 2 hours per 1000 at £6.10   12.20

Cost for 1000 slates ready for fixing   743.87

Materials: fixing slates and battens
Nr slates of width 255 mm per 100 m eaves: 100/0.255 = 392.16
Cost of slates per 100 m eaves: £743.87 × (392.16/1000)   291.71

Additional row of battens: 100 m at £0.28/m   28.00

Nr nails = (length battens)/(rafter centres) = 100/0.45
Weight of nails at 260 nails/kg: (100/0.45)/260 = 0.8547 kg
Cost of nails: 0.8547 kg at £2.50/kg   2.14

Labour: fixing slates and battens
Area slates per 100 m of eaves: 392.16/25.72 = 15.25 m^2
Labour cost per hour: (2 × £6.10) + (1 × £5.30) = £17.50
Output per hour per tradesman: 2.5 m^2
Cost to fix 100 m slates at eaves: £17.50 × 15.25/(2 × 2.5)   53.37

Squad of 2 tradesmen and 1 labourer fixes 80 m of battens per hour
Cost to fix 100 m battens: £17.50 × (100/80)   21.88
                                                                                        397.10
Profit and oncost 20%   79.42

Cost per 100 m   476.52

Rate per metre   £4.77

Example 9.4

> Verges: single extra undercloak course 115 mm wide; bedding and
> pointing in cement mortar 1:3.   m

Materials
Slate-and-a-half slates to alternate courses to form verge:
Heather blue slate ex quarry at £1525 per 1000   £1525.00

Assume site 200 miles from quarry
500 slates/crate weighing 1.00 tonnes/crate: by 16 tonne lorry
Nr of crates in load: 16/1.00 = 16
Nr of slates in load: 16 × 500 = 8000

Loading lorry:	1.12		
Unloading at site:	1.12		
Travelling time 1 man:	10		
Rest time etc.:	8		
	20.24 hours at £5.51	£111.52	
Lorry cost: 20.24 hours at £7.50		151.80	
Delivery cost of full load 8000 slates		263.32	
Delivery cost for 1000 slates: £263.32/8			32.92
Take delivery and stack: 1.5 hours per 1000 at £5.30			7.95
Nails: 40 mm copper at £5.50/kg at 2.7 kg per 1000: 2000 nails per 1000 slates			29.70

Labour: preparing slates

Holing slates per 1000: 4 hours at £6.10	24.40
	1619.97
Allow for waste 5%	81.00
Cost per 1000 slates prepared ready for fixing	1700.97

Material: fixing slates
Nr slates of gauge 153 mm per 100 m verge: $100/0.153 = 654$
Slate-and-a-half slate are used on alternate courses: $654/2 = 327$

Cost of slate-and-a-half slates: 327 at £1700.97 per 1000	556.22
Cost of 405×255 mm slates: 327 at £731.67 per 1000	239.26
Cost of slates per 100 m verge	795.48

Deduct cost of slates measured in superficial item (Example 9.2)

Coverage slate-and-a-half slates: $327 \times (0.153 \times 0.380)$	$19.01 \, m^2$	
Coverage 405×255 mm slates: $327 \times (0.153 \times 0.255)$	$12.76 \, m^2$	
Area covered by verge	$31.77 \, m^2$	
Cost to deduct: $31.77 \, m^2$ at £18.81 per m^2 (Example 9.2)		597.59
Extra over cost of slates per 100 m verge		197.89

Cement mortar 1:3: cost say £59.62 per m^3 (Appendix Figure A2)
Mortar required to bed 100 m verge: $100 \times 0.115 \times 0.025 = 0.2875$

Cost of mortar: $£59.62 \times 0.2875$	17.14
Slate undercloak 115 mm wide: 100 m at £1.50/m	150.00

Labour: fixing slates

Forming square verges including bedding: 14.5 hours per 100 m at £6.10	88.45
	453.48
Profit and oncost 20%	90.69
Cost per 100 m	544.17
Rate per metre	£5.44

Example 9.5

Hips: raking; on 405×255 mm heather blue slates.	m

Using trigonometry we can establish the plane angle at the hip for a particular roof. From this we can calculate the length of the cut per slate, and thereafter the number of slates per metre which will be used in wastage calculations.

As an example, take a roof of 30 degrees pitch with a standard hipped end, spanning 8 m (Figure 9.1). First,

$$\tan 30° = \frac{\text{height of rise of roof}}{\text{half span of roof}}$$

$$\text{height of rise} = 4 \times \tan 30° = 2.31 \text{ m}$$

Next,

$$\text{rafter length} = \sqrt{(2.31^2 + 4^2)} = 4.62 \text{ m}$$

Denote the plane angle at the hip as α. Then

$$\tan \alpha = \frac{\text{rafter length}}{\text{half span of roof}} = \frac{4.62}{4} = 1.155$$

$$\alpha = 49.11°$$

Now consider the cut slates:

$$\tan \alpha = \frac{\text{height of cut up longer edge of slate}}{\text{width of slate}}$$

$$\text{height of cut} = 1.155 \times 255 = 293 \text{ mm}$$

Then

$$\text{cut length} = \sqrt{(255^2 + 293^2)} = 389 \text{ mm}$$

Having calculated the length of cut on each slate, we must now determine the number of slates to be cut over a metre length of hip. This is done by first calculating the length of cut on the margin of each slate.

With a plane angle of 49.11 degrees and a margin of 153 mm (as the gauge previously calculated), we first determine the length of cut on the margin as follows:

$$\sin \alpha = \frac{\text{margin}}{\text{cut on margin}}$$

$$\text{cut on margin} = \frac{\text{margin}}{\sin \alpha} = \frac{153}{0.757} = 202 \text{ mm}$$

We can now move on to the estimation of costs.

Materials: wastage
Nr courses slate per metre hip: $1000/202 = 4.95$
Wastage on side of hip per metre: $4.95 \times 50\% = 2.48$ slates

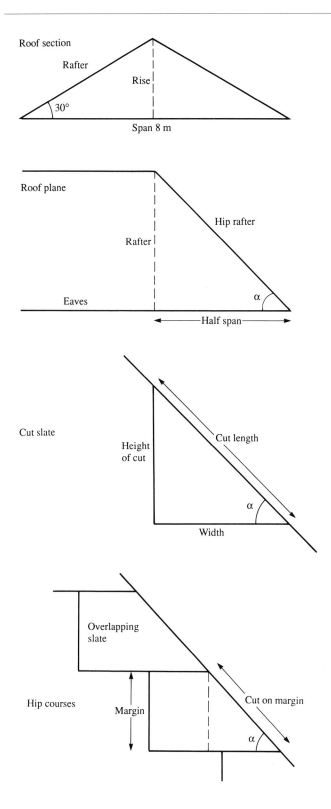

Figure 9.1 *Roof and slate angles*

Cost of 1000 slates from Example 9.2		
Ex quarry	£610.00	
Delivery to site	24.78	
Take delivery and stack	7.95	
	£642.73	
Cost of slates per metre hip: £642.73 × (2.48/1000)		£1.59

Labour: cutting
Cost squad per hour: $(2 \times £6.10) + (1 \times £5.30) = £17.50$
Output per tradesman: 10 m hip cut per hour

Cost to cut one metre: £17.50/(2 × 10)	0.88
Cost to cut one side of hip	2.47
Cost to cut two sides of hip	4.94
Profit and oncost 20%	0.99
Cost per metre	£5.93

Tile roofing

Roof tiles are manufactured from clay and concrete to a wide range of colours and profiles suitable for pitches from 17 to 45 degrees. Greater pitches up to vertical cladding are achievable but require additional or special fixing. Roof tiles are generally sold by the thousand in crates or strap banded to be stored on site pallets.

Clay plain roof tiles

British Standard BS 402 provides for tiles of nominal or work size 265 × 165 mm. Table 1 of the Standard gives maximum and minimum deviations for manufacturing size. Also detailed are requirements for tile manufacture, colour, nibs, thickness etc. Where tiles 280 × 175 mm are required for a project, but otherwise comply with BS 402, then such a requirement should be stated when ordering. BS 402 recognizes that plain tiles of different specification are manufactured, and although the Standard is not intended to provide for such tiles it does not rule out their use.

Plain tiles are laid using the double lap principle; this ensures that there will be at least two thicknesses of tile covering any part of the roof.

Concrete roofing tiles

In accordance with the requirements of BS 473,550, concrete roofing tiles or concrete roofing slates are manufactured to be fixed either double lap or single lap. Single-lap tiles are fitted using interlocking principles; the lower edge of a tile is designed to fit closely to the upper edge of the tile adjacent. The overall work sizes and maximum and minimum deviations for manufacturing size are given in the Standard. The manufacture of tiles in sizes other than those stated in the Standard is not excluded, and such tiles may therefore be found in practice.

Table 9.4 shows typical overall work sizes for single-lap concrete tiles, and an indication of labour outputs based on a squad of two tradesmen to one labourer.

Table 9.4 *Range of tiler outputs fixing single-lap concrete tiles*

Overall work size (mm)		Output per tradesman
Length	Width	(m²/hour)
381	229	3.50–4.00
413	330	3.25–3.75
420	330	3.25–3.75
430	380	3.00–3.50

Concrete double-lap tiles have similar characteristics to those of clay plain tiles. The work sizes given in the Standard are 267×165 mm and 457×330 mm, and maximum and minimum deviations for manufacturing size are again specified.

Example 9.6

> Roofing: concrete interlocking tiles 430×380 mm; fixing every tile with standard tile clip to 75 mm lap; battens 38×25 mm softwood sawn, pressure impregnated with preservative, on reinforced underslating felt; covering sloping to 40 degree pitch.   m²

Materials: tiles

Tiles delivered to site at £490 per 1000		£490.00
Take delivery and stack: 1.75 hours per 1000 at £5.30		9.27
Clips: 1000 per 1000 tiles at £25 per 1000		25.00
		524.27
Allow for waste 2.5%		13.11
Cost per 1000 tiles ready for fixing		537.38

Coverage 1 tile = gauge × exposed width
$$= (\text{length} - 75 \text{ mm lap}) \times (\text{width} - 35 \text{ mm lap})$$
$$= 0.355 \times 0.345 = 0.12247 \text{ m}^2$$
Coverage 1000 tiles: $0.12247 \times 1000 = 122.47 \text{ m}^2$
Cost per m²: £537/122.47   4.39

Underslating felt as Example 9.1: per m²   1.28

Materials: battens
Nr tiles per m²: $1000/122.47 = 8.16$
Quantity of 38×25 mm battens per m² = nr tiles per m² × exposed width
$$= 8.16 \times 0.345 = 2.82 \text{ m}$$
Cost of battens per m²: 2.82 m at £0.28 per metre   £0.79

Nr nails per m² = (length battens)/(rafter centres) $= 2.82/0.45$
Weight of nails per m² at 260 nails/kg: $(2.82/0.45)/260 = 0.024$ kg
Cost of nails per m²: 0.024 kg at £2.50/kg   0.06

		0.85
Allow for waste 5%	0.04	0.89

Labour: battens
2 tradesmen and 1 labourer fixes 80 m battens per hour
Labour cost per hour: $(2 \times £6.10) + (1 \times £5.30) = £17.50$

Cost to fix battens for 1 m² tiling: $£17.50 \times (2.82/80)$ 0.62

Labour: tiles
From Table 9.4, for the tile size of 430×380 mm, 3.5 m² can be carried and fixed in position per hour per tradesman. There are 2 tradesmen in the squad.
Labour cost per hour = £17.50
Cost per m²: = $£17.50/(2 \times 3.5)$ 2.50

    9.68
Profit and oncost 20%   1.94
Rate per m²   £11.62

Example 9.7

> Roofing: plain clay rosemary tiles 265×165 mm; fixing each tile with one aluminium alloy nail, every fifth course twice nailed, to 65 mm lap; battens 38×25 mm softwood sawn, pressure impregnated with preservative, on reinforced underslating felt; covering sloping to 40 degree pitch.   m²

Materials: tiles
Tiles delivered to site at £465 per 1000   £465.00
Take delivery and stack: 1.3 hours per 1000 at £5.30   6.89
Nails: 38 mm aluminium at £5.50/kg at 1.75 kg per 1000: 1200 per 1000 tiles   11.55

    483.44
Allow for waste 2.5%   12.09
Cost per 1000 tiles ready for fixing   495.53

Coverage 1 tile = width × gauge = $0.165 \times 0.1 = 0.0165$ m²
Coverage 1000 tiles = $0.0165 \times 1000 = 16.5$ m²
Cost per m² = £495.53/16.5   30.03

Underslating felt as Example 9.1: per m²   1.28

Materials: battens
Nr tiles per m²: $1000/16.5 = 60.61$
Quantity of 38×25 mm battens per m² = nr tiles per m² × tile width
    $= 60.61 \times 0.165 = 10$ m
Cost of battens per m² 10 m at £0.28 per metre   £2.80

Nr nails per m² = (length battens)/(rafter centres) = $10/0.6$
Weight nails per m² at 260 nails/kg: $(10/0.6)/260 = 0.064$ kg
Cost of nails per m²: 0.064 kg at £2.50/kg   0.16

    2.96
Allow for waste 5%   0.15   3.11

Labour: battens
Squad of tradesmen and 1 labourer fixes 80 m battens per hour
Labour cost per hour: $(2 \times £6.10) \times (1 \times £5.30) = £17.50$
Cost to fix battens for 1 m² tiling: $£17.50 \times (10/80)$ 2.19

Labour: tiles
For plain clay tiles 265×165 mm, 2.5 m² can be carried and fixed in position per hour per tradesman. There are 2 tradesmen in the squad.
Labour cost per hour: £17.50
Cost per m²: $£17.50/(2 \times 2.5)$ 3.50

    40.11
Profit and oncost 20%   8.02
Rate per m²   £48.13

Example 9.8

> Eaves: double course of 215×165 mm plain clay rosemary tiles at eaves.   m

Materials
Tiles delivered to site at £465 per 1000   £465.00
Take delivery and stack: 1.3 hours per 1000 at £5.30   6.89
Nails: 38 mm aluminium at £5.50/kg at 1.75 kg per 1000: 2000 per
    1000 tiles   19.25

    491.14
Allow for waste 2.5%   12.28
Cost per 1000 tiles ready for fixing   503.42

Nr tiles of width 165 mm per 100 m eaves: $100/0.165 = 606.06$
Cost tiles per 100 m eaves: $£503.42 \times (606.06/1000)$ 305.10

Additional row of battens: 100 m at £0.28/m   28.00

Nr nails = (length battens)/(rafter centres) = $100/0.6$
Weight of nails at 260 nails/kg: $(100/0.6)/260 = 0.64$ kg
Cost of nails: 0.64 kg at £2.50/kg   1.60

Labour: fixing battens and tiles
Squad of 2 tradesmen and 1 labourer fixes 80 m battens per hour
Labour cost per hour: $(2 \times £6.10) + (1 \times £5.30) = £17.50$
Cost to fix 100 m battens: $£17.50 \times (100/80)$ 21.87

Coverage 1 tile = width × gauge = $0.165 \times 0.10 = 0.0165$ m²
Coverage 1000 tiles = $0.0165 \times 1000 = 16.5$ m²
Nr tiles per m² = $1000/16.5 = 60.61$
Area tiles per 100 m eaves = (nr tiles per 100 m eaves)/(nr tiles/m²)
    $= 606.06/60.61 = 10$ m²

Labour cost £17.50/hour: output per tradesman 2.5 m²/hour
Cost to fix 100 m tiles at eaves: £17.50 × 10/(2 × 2.5)   35.00

	391.57
Profit and oncost 20%	78.31
Cost per 100 m	469.88
Rate per metre	£4.70

Example 9.9

> Verges: single extra undercloak course 165 mm wide; bedding and
> pointing in cement mortar 1:3.   m

Materials
Tile-and-a-half tiles delivered to site at £890 per 1000   £890.00
Take delivery and stack: 1.35 hours per 1000 at £5.30   7.15
Nails: 38 mm aluminium at £5.50/kg at 1.75 kg per 1000: 2000 per
    1000 tiles   19.25

	916.40
Allow for waste 2.5%	22.91
Cost per 1000 tiles ready for fixing	939.31

Nr tiles of gauge 100 mm per 100 m verge: 100/0.1 = 1000
Cost of tile-and-a-half tiles: 500 at £939.31 per 1000   469.66
Cost of 265 × 165 mm tiles: 500 at £495.53 per 1000   247.76

Cost of tiles per 100 mm verge   717.42

Deduct cost of tiles measured in superficial item (Example 9.7):
Coverage tile-and-a-half tiles: 500 × (0.1 × 0.248)   12.40 m²
Coverage 265 × 165 mm tiles: 500 × (0.1 × 0.165)   8.25 m²

Area covered by verge   20.65 m²

Cost to deduct: 20.65 m² at £30.03 per m² (Example 9.7)   620.12

Extra over cost of tiles per 100 m verge   97.30

Cement mortar 1:3: cost say £45 per m³
Mortar required to bed 100 m verge: 100 × 0.115 × 0.025 = 0.2875 m³
Cost of mortar: £45 × 0.2875   12.94

Plain tile undercloak 165 mm wide: 100 m at
    (£495.53/1000) × (1000/265) = £1.87/m   187.00

Labour
Forming square verges including bedding: 18.5 hours per 100 m at £6.10   112.85

	410.09
Profit and oncost 20%	82.02
Cost per 100 m	492.11
Rate per metre	£4.92

Example 9.10

> Ridge: plain half round clay tiles 305 × 200 mm dia. bedding and
> pointing in cement mortar 1:3.   m

Materials

Tiles delivered to site at £430 per 1000	£430.00
Take delivery and stack; 1.5 hours per 1000 at £5.30	7.95
	437.95
Allow for waste 2.5%	10.95
Cost per 1000 tiles ready for fixing	448.90

Cement mortar 1:3: cost say £45 per m^3
Mortar required to bed 1 half round tile, say 25% tile volume
$= \pi \times 0.1 \times 0.1 \times 0.305 \times 0.25 = 0.0024$ m^3
Mortar to bed 1000 tiles: 2.4 m^3

Cost of mortar: £45 × 2.4	108.00
Cost per 1000 ridge tiles	556.90

1000 ridge tiles cover (length + mortar joint) × 1000 = (305 + 10)mm × 1000
$= 315$ m

Cost per metre: £556.90/315	1.77

Labour
Tradesman fixes 4 m ridge tiles per hour. Squad contains 2 tradesmen and 1 labourer
Labour cost per hour: (2 × £6.10) + (1 × £5.30) = £17.50

Cost per metre: £17.50/(2 × 4)	2.19
	3.96
Profit and oncost 20%	0.79
Rate per metre	£4.75

Lead sheet coverings and flashings

Milled lead sheet is manufactured to BS 1178:1969 'Milled lead sheet and strip for building purposes'. A range of six sizes of lead sheet, defined by thickness in millimetres, is provided for in this Standard (see Table 9.5). A colour coding system is used to enable easy identification of thicknesses.

Lead sheet is supplied by the manufacturer cut to the dimensions required, or in large sheets, the largest sheet size being 2.4 m wide by up to 12 m long.

Lead strip is defined as cut to widths from 75 to 600 mm. Lead strip is supplied in coils and is the most common form of lead sheet used, being convenient for flashing and weathering applications.

Fixing accessories for external leadwork

The following accessories are normally associated with the fixing of lead sheet in the traditional manner.

Table 9.5   *Lead sheet sizes*

BS code	Thickness (mm)	Weight (kg/m^2)	Colour code
3	1.25	14.18	Green
4	1.80	20.41	Blue
5	2.24	25.40	Red
6	2.50	18.36	Black
7	3.15	35.72	White
8	3.55	40.26	Orange

Copper clips

Copper clips should be cut from copper sheet not less than 0.6 mm thick and conform with BS 2870.

Nails

Nails should be copper clout nails with jagged shanks conforming to BS 1202 Part 2, Table 2. They should not be less than 25 mm long or 10 SWG shank diameter.

Screws

Screws should be of brass or stainless steel to BS 1210. They should be not less than 25 mm long or 10 SWG.

Solder

Solder for making fixing dots, angles and other seams should conform with BS 219 grade D or grade J.

Underlay

The underlay used to isolate lead roofing, cladding and gutter linings from their substrates is impregnated felt or waterproof building paper. Impregnated felt should be used where the surface is not perfectly even. Where the surface is even, waterproof building paper to BS 1521 class A would be suitable.

Outputs

Ranges of squad outputs for lead sheet coverings are given in Table 9.6. The outputs include hoisting, cutting and fixing lead per square metre as measured.

Ranges of squad outputs for lead flashings are given in Table 9.7. The outputs include hoisting, cutting and fixing lead per linear metre as measured.

Table 9.6 *Outputs in m²/hour for fixing lead sheet covering: squad of plumber and mate*

BS code	Flat roofs: <10° from horizontal	Roofs 10° – 50° from horizontal	Roofs 50° from horizontal to vertical
3	—	—	—
4*	0.90–1.00	1.00–1.10	1.40–1.55
5	1.15–1.25	1.30–1.40	1.70–1.80
6	1.35–1.45	1.50–1.60	2.05–2.20
7	1.55–1.65	1.70–1.85	2.30–2.45
8	1.80–1.90	2.00–2.20	2.60–2.80

*Code 4 lead sheet is suitable only for coverings to small areas for the types of work shown.

Tale 9.7 *Outputs in metres/hour for fixing lead sheet flashings: squad of plumber and mate*

BS code	Description	Girth (mm) <150	150–225	225–300
4	Flashing	0.25–0.30	0.35–0.40	0.40–0.45
5	Flashing	0.30–0.35	0.40–0.45	0.50–0.55
4	Stepped flashing	0.50–0.55	0.65–0.70	0.80–0.90
5	Stepped flashing	0.60–0.65	0.80–0.90	1.00–1.15
4	Hip or ridge capping	0.30–0.36	0.40–0.45	0.50–0.60
5	Hip or ridge capping	0.35–0.40	0.50–0.55	0.60–0.65
4	Sloping gutters	0.30–0.35	0.40–0.45	0.50–0.55
5	Sloping gutters	0.35–0.40	0.50–0.55	0.60–0.65
6	Sloping gutters	0.40–0.45	0.60–0.65	0.70–0.75

Example 9.11

> Roof coverings: flat; code 5 lead sheet on Erskine's felt underlay, fixing at head with copper clout nails 25 mm long 10 SWG, copper side edge and end retaining clips; wood core rolls at 1000 mm centres; drips every 2000 mm.   m²

Materials

Cost of 1000 kg of lead sheet	£780.00
Plumber's mate taking delivery at 1000 kg/hour at £5.30	5.30
Cost per 1000 kg	785.30

Cost per m² of code 5: £785.30 × (25.4/1000)   £19.95

Additional 250 mm width of lead required forming laps over 45 mm wood roll:
Cost for 250 mm lap per m²: £19.95 × (250/1000)   4.99

Additional 180 mm length of lead required forming drips every 2 m:
Cost for 180 mm drip per m²: £19.95 × 180/(2 × 1000)   1.80

Copper clout nails per m²: say 31 at 120/kg at £3.50/kg   0.90
Copper clips at 500 mm centres fixed with 2 nails/clip (nails allowed for above):
3 clips per m² at £0.03   0.09
Cost per m² for underlay including allowance for laps and drips: say   1.15
                                                                                    _____
                                                                                    28.88
Allow for waste 2.5%   0.72

Labour
Squad will hoist, cut and fix 1 m² in 1.2 hours:
Plumber: 1.2 hours at £7.85*   9.42
Plumber's mate: 1.2 hours at £7.45*   8.94
                                                                                    _____
                                                                                    47.96
Profit and oncost 20%   9.95

Rate per m²   £57.55

*See Chapter 14.

Example 9.12

Flashings: horizontal; code 4 lead sheet; turning into groove and fixing with milled lead wedges at 200 mm centres; 150 mm girth.	m

Materials
Cost of 1000 kg of lead sheet   £780.00
Plumber's mate taking delivery at 1000 kg/hour at £5.30   5.30
                                                                    _____
Cost per 1000 kg   785.30

Cost per m² of code 4: £785.30 × (20.41/1000)   £16.03

Cost per metre at 150 mm girth: £16.03 × 0.15   £2.40

Additional 100 mm length of lead required forming laps every 2 m:
Cost for 100 mm lap per metre: £2.40 × 100/(2 × 1000)   0.12

Wedges 5 per metre: say   0.12
                                                                                 _____
                                                                                 2.64
Allow for waste 2.5%   0.07

Labour
Squad will hoist, cut and fix 1 m in 0.28 hours:
Plumber: 0.28 hours at £7.85   2.20
Plumber's mate: 0.28 hours at £7.45   2.09
                                                                                 _____
                                                                                 7.00
Profit and oncost 20%   1.40

Rate per metre   £8.40

Example 9.13

> Gutters: sloping; code 5 lead sheet; dressing over tilting fillets; fixing with copper clout nails; 550 mm girth.　　　　　　　　　　　　m

Materials
Cost of 1000 kg of lead sheet	£780.00	
Plumber's mate taking delivery at 1000 kg/hour at £5.30	5.30	
Cost per 1000 kg	785.30	
Cost per m³ of code 5: £785.30 × (25.4/1000)	£19.95	
Cost per metre at 550 mm girth: £19.95 × 0.55		£10.97
Additional 100 mm length of lead required forming laps every 2 m: Cost for 100 mm lap per metre: £10.97 × 100/(2 × 1000)		0.55
Copper clout nails per metre: say 15 at 120/kg at £3.50/kg		0.44
		11.96
Allow for waste 2.5%		0.30

Labour
A squad will hoist, cut and fix 1 m in 0.73 hours:		
Plumber:　　　0.73 hours at £7.85		5.73
Plumber's mate: 0.73 hours at £7.45		5.44
		23.43
Profit and oncost 20%		4.69
Rate per metre		£28.12

Example 9.14

> Edges: welted.　　　　　　　　　　　　　　　　　　　　m

Plumber forms 1 m of welted edge on code 5 lead in 0.12 hours:	
Plumber 0.12 hours at £7.85	£0.94
	0.94
Profit and oncost 20%	0.19
Rate per metre	£1.13

10

Asphalt work
SMM7 Sections J20, J21, M11

Introduction

It is perhaps unfair to refer to any particular trade as special, for every operative in the industry demonstrates skills which are unique to the craft he follows. However, asphalt work is generally referred to as a specialist trade because it encompasses some unique features and is practised by a relatively small number of craftsmen employed by a relatively small number of companies. To further emphasize this specialism, contracts for this type of work are normally let as a subcontract and most commonly to a nominated subcontractor.

The most unique aspect of the trade is the material used. Unlike every other trade, the material – mastic asphalt – is handled by the operatives while it is hot, generally between 200 and 220°C. Pitch mastic is laid at the lower temperature of 160°C.

This has two consequences. The first is that the work is dangerous; protective clothing is essential and safety procedures must be adhered to.

The second is that a relatively rapid pace is needed in the application of the material, as it sets by cooling to ambient temperature. As soon as the material is drawn from the cauldron or mixer it starts to cool. When it is applied to a surface the rate of cooling accelerates for two reasons. First, the base surface is at ambient temperature – and a base of masonry presents an enormous heat sink. Second, as the material is spread out, more of it comes into contact with this cold base and with the cool air over it.

Perhaps as a consequence of the specialist nature of the work there seems to be no well defined procedure used by estimators for building up rates for measured items of work. Therefore the data and examples which follow are given only as a guide to one possible approach to achieve comprehensive financial cover for all expenditure.

Before commencing examples it is proper to consider the logistics and practice in the trade.

Plant

Plant can be divided between melting equipment and small tools.

Melting equipment comes in a range of capacities, that is the weight of asphalt or pitch with which it is possible to deal in one operation. The smaller capacity items are called cauldrons or pots. Pots comprise a cylindrical lidded container with a heat source underneath. These pots are generally fired from cylinders of propane gas. While mastic asphalt is being melted and kept molten it is occasionally stirred to keep the temperature even. Pitch mastic has to be continuously stirred, and pots for use with that material have mechanically driven paddles built in for that purpose. Pots generally have a capacity of 500 kg of asphalt; this will take approximately 25 kg of propane to melt down.

Larger capacity plants are called mixers. The melting chamber incorporates paddles for mechanical mixing. The fuel used was once coke or coal but this has been replaced with oil or propane. Mixers may be mounted permanently on either a lorry or a trailer chassis. Usually mixers have a capacity of 1500 kg of asphalt, and naturally take proportionately more fuel to melt down all the asphalt.

Small tools and accessories include the following:

- Various trowels for spreading asphalt, working fillets, turning into grooves etc.
- Lidded buckets for transporting the melted asphalt from pot to site of laying
- Rules, straight edges and gauges
- Tin snips, pliers, hammers, knives etc. to deal with underlays and reinforcing meshes

Materials

Rock and mastic asphalt and pitch mastic are purchased by weight and supplied in blocks or cakes, usually weighing 25 kg.

Typically 1 tonne (1000 kg) of asphalt will cover 35 m^2 to a thickness of 12 mm, this being the thinnest single coat usually laid in one-coat work. Coats as thin as 10 mm are laid but only in combination with other coats. This cover rate takes into account a certain degree of unevenness in the substrate. The figure of 12 mm therefore would be used for a first coat. Subsequent coats will have an increased cover rate as there is no need to take any unevenness into account; the increase in cover rate is approximately 5 per cent.

So for a layer 20 mm thick, whatever the number of coats, the quantity of asphalt used per m^2 would be as follows:

First 12 mm thickness: 1000/35	28.57 kg
Further 8 mm thickness: $(1000/35) \times (8/12) \times 1.05$	20.00 kg
	48.57 kg

This is equivalent to $1000/48.57 = 20.59$ m^2 per tonne. Waste is not included in the calculation; it is normally added to the cost of the material when working out the rate (see later).

Bituminous felt and glass fibre cloth or tissue are used where an underlay is required; the bituminous felt should be type 4A(i) complying with BS 747. Bitumen coated expanded metal lath is used as a reinforcement for vertical work and in upstands, where adhesion to the substrate is not possible; this is supplied in sheets of approximately 2500×700 mm. Galvanized clout nails or wire staples are used to secure underlay and reinforcement to timber substrates.

Rubber/bitumen primer is used to coat metallic surfaces prior to the application of asphalt. This ensures a watertight bond. Coverage is approximately 10 m^2 per litre. Surfaces treated might include lead flashings and drips, cast iron rainwater outlets and gullies in roofs and pavings, pipes passing through a surface, balusters and other metal work. The primer has a solvent base – usually petroleum spirit – and is highly flammable in its liquid state.

Sundry materials used include crushed limestone dust to line buckets, thus preventing the asphalt or pitch mastic sticking to the bucket, and sharp sand to obtain a sand rubbed finish.

Labour

The size and nature of the job, the timetable and the accessibility of the works will determine whether pots or mixers are to be used. There may be a combination of a mixer for the large areas and a pot for the ancillary work such as skirtings, upstands, kerbs and drips.

The squad make-up and cost are generally broken down as follows:

2 craftsmen (spreaders) at £6.10	£12.20
1 potman at £5.90	5.90
1 labourer at £5.30	5.30
Squad cost per hour	23.40
Cost per craftsman per hour	£11.70

The output for the squad is based on how much material a craftsman can lay per squad hour. The output of the craftsman is affected by the nature of the work and its location. For example, work may be in a basement, in an underpinning, on a roof or on a floor; it may be in small areas or confined spaces; it may be sloping or vertical; and it may require to be done overhand. Typical outputs are given in Table 10.1.

Rule 1.*.*.3 of SMM7 Section J 'Waterproofing' is generally only applicable to cold applied materials.

Table 10.1   *Range of outputs per craftsman for mastic asphalt work*

(a) Area outputs: m²/hour

Thickness (mm)	12–13	15–17	18–20
Coats	1	2	2
Laying flooring/paving	0.24–0.43	0.29–0.46	0.32–0.49
Laying roofing:			
Horizontal	0.3 –0.6	0.33–0.75	0.5 –0.9
Sloping	0.4 –0.72	0.6 –1.08	0.9 –1.38
Vertical	1.0 –1.56	1.2 –2.16	1.85–2.4

(b) Linear outputs: m/hour   Enumerated outputs hours/each

Skirtings, aprons, upstands: girth 150 mm	0.36–1.00
225 mm	0.44–1.32
300 mm	0.5 –1.65
Angles: girth 150 mm	0.45–0.6
225 mm	0.65–0.8
300 mm	0.85–1.00
Forming channels in asphalt: girth 225 mm	0.7 –0.9
300 mm	1.0 –1.2
450 mm	1.25–1.40
Rounded edges	0.2 –0.33
Arris: horizontal	0.3 –0.5
vertical	0.25–0.4
Internal angle fillet: horizontal	0.15
vertical	0.2
Working into groove: horizontal	0.3
vertical	0.35

Wastage

Generally 2.5–5 per cent is allowed on asphalt and underlay. Small amounts of reinforcement and the priming of numerous small items will incur a greater degree of wastage in those particular items. However, the difference is small enough to be ignored and a general 5 per cent is used throughout the examples here.

Further quantities for examples

Before commencing examples of building up rates, let us establish a charge for the major plant item – the pot – and then set a small scenario within which to price a few items.

The hire rate for a half tonne pot for one week is £34.00. For a 39 hour week, the hire is therefore £0.87 per hour.

The total area of asphalt to be laid is taken to be 230 m² plus sundry skirtings, angle fillets, drips etc. Generally the thickness is 20 mm. Thus, using the figure calculated earlier, the approximate quantity of asphalt is 230 × 48.57 kg, plus 10% for other items: say 12.29 tonnes.

We will take 50 kg propane gas to cost £31.58.

Example 10.1

20 mm thick mastic asphalt in two coats to horizontal concrete substrate, including isolating membrane.	m²

Material: 48.57 kg at £85/tonne	£4.13
Type 4A(i) sheathing felt as isolating membrane:	
Roll 20 m² costs £10: divide by 19 m² to allow for laps	0.53
	4.66
Waste 5%	0.23
	4.89
Plant:	
500 kg asphalt is melted in 1.5 hours	
To melt 48.57 kg: hire charge (£0.87 × 1.5 × 48.57)/500	0.13
Transport pot to and from site: £25 shared over 12.29 tonnes	
So for 48.57 kg say (£25.00 × 48.57)/12.29 × 1000)	0.10
Propane to melt 500 kg costs £31.58	
So to melt 48.57 kg will cost (£31.58 × 48.57)/500	3.07
	8.19
Labour laying 3 m²/hour at £11.70	3.90
Pot hire charge while laying £0.87/hour: £0.87/3	0.29
	12.38
Profit and oncost 20%	2.48
Rate per m²	£14.86

Example 10.2

20 mm thick mastic asphalt in two coats to horizontal timber substrate, including isolating membrane nailed to timber deck.	m^2

Material: 48.57 kg at £85/tonne	£4.13
Type 4A(i) sheathing felt as isolating membrane:	
Roll 20 m^2 costs £10: divide by 19 m^2 to allow for laps	0.53
Clout nails 150 mm centres each way: 0.10 kg at £2.85	0.28
	4.94
Waste 5%	0.25
	5.19
500 kg asphalt is melted in 1.5 hours	
To melt 48.57 kg: hire charge at £0.87/hour	0.13
Transport pot to and from site: £25 shared over 12.29 tonnes	
So for 48.57 kg say	0.10
Propane to melt 500 kg costs £31.58	
So to melt 48.57 kg will cost	3.07
	8.49
Labour laying 3 m^2/hour at £11.70	3.90
Pot hire charge while laying £0.87 per hour: £0.87/3	0.29
	12.68
Profit and oncost 20%	2.54
Rate per m^2	£15.22

Example 10.3

20 mm thick mastic asphalt in two coats to vertical face of brickwork.	m^2

Material: 48.57 kg at £85/tonne	£4.13
Waste 5%	0.21
	4.34
500 kg asphalt is melted in 1.5 hours	
To melt 48.57 kg: hire charge at £0.87/hour	0.13
Transport pot to and from site: £25 shared over 12.29 tonnes	
So for 48.57 kg say	0.10
Propane to melt 500 kg costs £31.58	
So to melt 48.57 kg will cost	3.07
	7.64
Labour laying 0.4 m^2/hour at £11.70	29.25
Pot hire charge while laying £0.87 per hour: £0.87/0.4	2.17
	39.06
Profit and oncost 20%	7.81
Rate per m^2	£46.87

Example 10.4

20 mm thick mastic asphalt in two coats in skirting 150 mm high, including angle fillet, rounded arris and tucking into groove in brickwork.	m

Volume of material in skirting (m³/m):

Main upstand 150 × 20 mm	0.0030
Angle fillet 50 mm wide across angle: (50 × sin 45°) = 35.36 mm side	0.0006
Groove in brickwork	0.0005
	0.0041

Density of asphalt 2381 kg/m³

Material: 2381 × 0.0041 = 9.76 kg at £85/tonne	£0.83
Waste 5%	0.04
	0.87

500 kg asphalt is melted in 1.5 hours

To melt 9.76 kg: hire charge	0.03
Transport pot to and from site: £25 shared over 12.29 tonnes	
So for 9.76 kg say	0.02
Propane to melt 500 kg costs £31.58	
So to melt 9.76 kg will cost	0.62
	1.54
Labour laying 2 m/hour at £11.70	5.85
Pot hire charge while laying £0.87 per hour: £0.87/2	0.43
	7.82
Profit and oncost 20%	1.56
Rate per metre	£9.38

Example 10.5

20 mm thick mastic asphalt in two coats in skirting 150 mm high, on free-standing timber kerb, including angle fillet, reinforcement and isolating membrane.	m

Volume of material in skirting (m³/m)

Main upstand 160 × 20 mm (160 girth over timber)	0.0032
Angle fillet 50 mm wide across angle: (50 × sin 45°) = 35.36 mm side	0.0006
	0.0038

Density of asphalt 2381 kg/m³

Material: 2381 × 0.0038 = 9.05 kg at £85/tonne	£0.77
Reinforcement: 225 mm wide at £1.04/m²	0.23
Type 4A(i) sheathing felt: 150 mm wide at £0.53/m²	0.08
Nails, say	0.02
	1.10
Waste 5%	0.06
	1.16

500 kg asphalt is melted in 1.5 hours	0.02
To melt 9.05 kg: hire charge at £0.87/hour	0.02
Transport pot to and from site: £25 shared over 12.29 tonnes	
So for 9.05 kg say	0.02
Propane to melt 500 kg costs £31.58	
So to melt 9.05 kg will cost	0.57
	1.77
Labour laying 1.8 m/hour at £11.70	6.50
Pot hire charge while laying £0.87 per hour: £0.87/1.8	0.48
	8.75
Profit and oncost 20%	1.75
Rate per metre	£10.50

Example 10.6

Internal angle fillet.	m

Volume of material in fillet (m^3):	
Angle fillet 50 mm wide aross angles: $(50 \times \sin 45°) = 35.36$ mm side	0.0006
	0.0006
Density of asphalt 2381 kg/m^3	
Material: $2381 \times 0.0006 = 1.43$ kg at £85/tonne	£0.12
Waste 5%	0.01
	0.13
500 kg asphalt is melted in 1.5 hours	
To melt 1.43 kg: hire charge at £0.87/hour	0.004
Transport pot to and from site: £25 shared over 12.29 tonnes	
So for 1.43 kg say	0.003
Propane to melt 500 kg costs £31.58	
So to melt 1.43 kg will cost	0.09
	0.23
Labour laying 6 m/hour at £11.70	1.95
Pot hire charge while laying £0.87 per hour: £0.87/6	0.15
	2.33
Profit and oncost 20%	0.46
Rate per metre	£2.79

Example 10.7

Forming channel girth 300 mm in asphalt.	m

Labour forming 1 m/hour at £11.70	£11.70
Profit and oncost 20%	2.34
Rate per metre	£14.04

11

Woodwork carpentry: first fixings

SMM7 Sections G20, K10, K31

The rationale behind the layout and structure of the rules of measurement in SMM7 has done away with the work section/trade headings which have been so familiar in the past. The versatility of wood and wood products and of the carpenters and joiners themselves now means that the rules of measurement for their work are well scattered through the new document. However, in this and the following chapter a selection of items and rates for these traditional trade divisions is presented in an order similar to that of SMM7.

Special requirements

In dealing with any woodworking items the estimator will be on the lookout for the following information, which will denote something additional or unusual – and therefore possibly expensive!

Species of timber

Hardwoods are generally more expensive than softwoods, and in either classification there is a wide range of cost due to what might appear to be a diversity of factors. These however generally all boil down to scarcity of a suitable supply of timber. It is also true to say that timbers are more expensive when required in larger cross-sections and/or long continuous lengths. These latter factors can be overcome using machine jointing and modern synthetic glues; the joint is often stronger than the raw timber but the jointing process does add to the cost. However, if timbers in excess of 5.4 m long are required this will add around 12.5 per cent to the basic cost of timber. In excess of 7.5 m long an addition of 25 per cent or more would be quite likely

Stress grading

This is an added expense but one which may save money in the long run by allowing the use of smaller sections of timber. Currently stress grading adds from around 7.5 per cent for GS grade timber to around 20 per cent for SS grade timber to the basic cost of timber.

Grade of sheet material

There are many different board and sheet materials on the market. A large number of them might be described as, say, plywood. However, there is a vast difference in price between a Douglas fir sheathing plywood and a hardwood veneered plywood for decorative work, or between one where the glue is graded water and boil proof (WBP) and another graded moisture resistant (MR).

Similarly, there are many different types and grades of particle board, and so on.

Pre-treatment

This covers treatment with chemicals against insect or fungal attack or for increased fire resistance. The treatment may be applied with a simple brush, in vacuum/pressure plant etc. Note that sheet material can be pre-treated as well as natural timber. Currently pressure impregnation against fungal or insect attack with copper–chrome–arsenate (CCA) salts based preservative adds around 12.5 per cent to the basic cost of timber.

Machined timber

Timber may be supplied regularized or wrot rather than off the saw, or may be sawn to a special shape.

It is impossible to say what percentage to add to the basic timber cost for regularizing or machine planing, since the act of machining the surface of a rectangular shape is much the same whether the cross-section is 38×38 mm or 100×100 mm. What does vary of course is the proportion of waste produced to the remaining processed timber. For a section 38×38 mm (1444 mm^2), fully wrot surfaces will mean that the timber will end up as 32×32 mm (1024 mm^2) maximum. This represents a loss of 420 mm^2 or 29.09 per cent. The 100×100 mm timber ($10,000$ mm^2) will end up as 94×94 mm (8836 mm^2). This represents a loss of 1164 mm^2 or 11.64 per cent. If the timber is 200×200 mm, the loss will be 5.91 per cent. The following are typical costs of these timbers as quoted in the construction magazines current at the end of 1990:

Basic size (mm)	Cost per 100 m	
	Sawn	Wrot
38×38	£35	£57
100×100	£217	£290
200×200	£810	£870

Examples of special shapes produced by sawing are arris, cant and firring. These add around 15 per cent to the cost of the basic timber.

Timber sizes and costs

The raw material, whether for direct use on site or for prefabrication, may be purchased by the metre. Usually the price quoted is for 100 m of a specific cross-section. More rarely timber is purchased by the cubic metre, but generally this is by the shipload and without conversion or processing between unloading and delivery.

Timber is supplied in a range of cross-sections. The following common names are approximately in descending order of cross-sectional area: baulks, half-timbers, flitches, planks, deals, boards, strips, battens, squares, slating battens and scantlings.

Timber is also supplied in one of four conditions: sawn or off the saw; regularized or surfaced, i.e. machined to a constant cross-section; wrot or dressed, i.e. machined to a constant cross-section and a smooth finish suitable for joinery work; and moulded, i.e. wrot and with a shaped cross-section.

BS 4471, 'Sizes of sawn and processed softwood', gives the sawn sizes of softwoods together with the reduction in overall size when the timbers are regularized or wrot. Note that flooring boards are treated differently and are the subject of BS 1297, 'Grading and sizing of softwood flooring'. Material from which UK sawmills machine flooring boards is specifically sawn for that purpose at the original sawmill before being shipped to the UK; the contractor buys the machined boards. BS 5450, 'Sizes of hardwoods', gives sawn sizes and processing allowances for planing etc.

As well as standard cross-sectional sizes, softwoods are supplied in standard lengths. The shortest length normally supplied is 1.8 m. Standard lengths increment by 300 mm from this basic length. The range of sizes continues therefore as 2.1, 2.4, 2.7, 3.0 m etc.

If these lengths do not fit the work required of them exactly, there is a maximum wastage of 299 mm to cut the next largest size down. The bulk of timber is purchased in the range of lengths from 3 to 6 m. If the wastage of 299 m is taken on the average length of 4.5, it is equivalent to 6.64 per cent. Many authorities quote 7.5 per cent allowance for wastage, which has also to include pilferage and damage. In the examples here we will take 10 per cent on timber, 7.5 per cent on manufactured units such as trusses, and 2.5 per cent on manufactured units such as doors and windows.

Although supplied in such a wide variety of sizes, natural timber is generally priced per cubic metre for a range of sizes and finishes. Pre-treatment is also costed out on a cubic metre basis. Prefabricated work (manufactured units) is generally purchased on a unit basis, i.e. per door, per flight of stairs, per kitchen unit, per trussed rafter etc.

As in all the other trades, the estimator must ensure that delivery costs to site are included in any quotes or must make an appropriate allowance. Unloading on site is most often done by having a mechanical off-load (MOL) facility on the delivery truck, or by having rough terrain fork-lifts to cope with banded or palleted materials. A fork-lift would be priced in the preliminaries section.

Fastenings and adhesives

Fastenings and adhesives are not generally measured separately. Nails are deemed to be included; all other fastenings are given in the item description. There is a wide range of fastenings available, each suited to a particular application.

Nails can be manufactured with round flat heads or bullet heads, and from round or oval wire. The materials range through mild steel, stainless steel, brass, copper, aluminium, galvanized steel, zinc plated steel, sherardized steel etc. Shanks can be plain or jagged, or can have an annular ring barb to improve holding power. An old form of nail which was punched out of steel sheet has enjoyed a revival in recent years, as it has been found suitable for direct fastening into autoclaved aerated concrete blockwork and no-fines concrete. These nails were known as cut nails and were produced in a variety of patterns, one of which was specifically designed for fastening flooring boards. Others were designed for anything from general carcassing to fine cabinet and upholstery work. Nails are well illustrated in BS 1202.

Wood screws are manufactured in a similar wide range of materials and finishes, and with a variety of heads (countersunk, round, pan etc. as well as slotted, Phillips, Posidrive etc.) and thread forms. Wood screws are fully specified in BS 1210. In addition there is a range of bolts, lag screws, fastenings for attachment to masonry, patent clips, hangers, hold-down straps etc., far too numerous to mention here. Manufacturers' catalogues give details of proprietary fastenings.

Glues can have several different prime sources, including animal bones, hooves and horns, fish, milk, cellulose and various synthetic rubbers. Glues based on animal, milk and cellulose products are generally classified as being from unmodified natural materials. Since the late 1940s there has been a rapid growth in products based mainly on synthetic materials, and these have provided the industry with adhesives which cover every requirement. BS 5442:Part 3 gives guidance on the suitability of all the glues mentioned above for use with wood. The synthetic materials are classified as synthetic thermoset resins, elastomers and thermoplastics.

The carpenter and joiner is expected to fix items of ironmongery ranging from simple hat and coat hooks through complex items securing doors and windows, patent fasteners and openers for sashes and doors, to control gear for large industrial doors. The range of ironmongery is too extensive to list here, and manufacturers' catalogues and instructions should be consulted.

Round wire nails in plain steel are the most common method of fastening for general carcassing work. Galvanized round wire nails are used when simple rusting would be a problem. Under the rules of measurement, plain steel nails are the norm and are deemed to be included. Galvanized nails have to be given in the item description, the bill headings or the preamble. Similar provisions apply for nails in other materials and finishes, and for other more complex (and more expensive) forms of fasteners.

Labour constants

In general terms, hardwood is more difficult to work with than softwood. Various authorities quote labour constants for softwood, and show multipliers ranging from 1.5 to 2.5 to be applied to these constants for working with hardwood in general. Others give multipliers for specified hardwoods, recognizing that not all hardwoods have the same physical hardness or ease of cutting and machining.

Of course neither are all softwoods the same. Some of the so-called softwoods are harder and heavier and more difficult to machine than the majority of hardwoods; for example, pitch pine (*Pinus palustris*) is quite dense and very resinous, which makes it difficult to work with hand tools. Modern construction practice tends to utilize pines, spruces and firs specially grown or selected from natural stands for their easy working capabilities and a general uniformity of quality. This obviates any need to differentiate between varieties as far as labour constants are concerned. The estimator must beware however of the occasional use of some unusual material in carpentry work. Many Far Eastern and South American exporters are promoting the use of unusual hardwoods for carpentry work – in the face of an environmental movement in Europe concerned about the felling of rain forests!

Table 11.1 gives a range of labour constants per linear metre for various carpentry timbers based on their cross-sectional area. Table 11.2 gives a further range of constants per cubic metre for general classifications of carpentry timber(s). Both tables assume general softwood – pines, spruces and firs.

Table 11.2 must be treated with caution. It is not feasible to convert these labour hours for 1 m^3 directly into hours per metre for a particular cross-section of joist, plate etc. Table 11.3 shows the results of such a conversion for sleeper joists using a figure of 20 hours/m^3. Comparison with the figures in Table 11.1 shows how the labour for floor timbers has been overestimated. Similarly, Table 11.4 shows calculations for roof timbers on the basis of 30 hours/m^3. Again these overestimate the constants per metre.

So why bother with hours for 1 m^3 of timber? Well, it is convenient for pricing works where there are no detailed quantities and no time to produce them. The inaccuracies will even out if the job is large enough, but the total result can only be approximate estimate and not an accurate prediction of cost. Therefore costings based on m^3 of timber must be treated with care. In the examples here, constants based on individual cross-sectional areas will be used.

Table 11.1 *Range of labour constants for fixing selected carcassing timbers (per carpenter)*

Item	Floor/roof joists/ties (hours/m)	Other roof timbers (hours/m)
Wall plates (any size)	0.04–0.13	0.04–0.13
Other: 100 × 50, 125 × 50 mm	0.05–0.17	0.10–0.18
150 × 50, 175 × 50 mm	0.07–0.17	0.14–0.20
200 × 50, 225 × 50 mm	0.10–0.19	0.18–0.24
Strutting: solid	0.33–0.40	
herringbone	0.30–0.50	
Nails		0.15–2.00 kg/m³
Trimming openings:		
Trim and house one timber		0.20–0.40 hours each
Tusk tenon and house to one timber		1.00–2.00 hours each

Table 11.2 *Labour constants for fixing 1 m³ of carcassing timbers (per carpenter)*

Item	hours/m³
Wall plate	7–12
Any floor or ceiling joist	8–20
Other roof timbers	20–30
Nails	1.1–4 kg/m³

Table 11.3 *Conversion of a labour constant of 20 hours/m³ to constants per metre for floor joists (per carpenter)*

First size (mm)		Second size (mm)	m/hour	hours/m
50	×	100	10.00	0.100
50	×	125	8.00	0.125
50	×	150	6.67	0.150
50	×	175	5.71	0.175
50	×	200	5.00	0.200
50	×	225	4.44	0.225

Table 11.4 *Conversion of a labour constant of 30 hours/m³ to constants per metre for roof timbers (per carpenter)*

First size (mm)		Second size (mm)	m/hour	hours/m
50	×	100	15.00	0.150
50	×	125	12.00	0.188
50	×	150	10.01	0.225
50	×	175	8.56	0.263
50	×	200	7.50	0.300
50	×	225	6.66	0.338

Further quantities for examples

It should be explained that the use of labourers by carpenters is not common and depends very much on the custom and size of the company in which they work. For the purposes of the following examples, a squad of five carpenters and one labourer is assumed. The squad cost is broken down as follows:

1 labourer at £5.30	£5.30
5 carpenters at £6.10	30.50
Squad cost per hour	35.80
Cost per carpenter hour	£7.16

It is further assumed that timber and components delivered to site are banded or palleted for mechanical off-loading. Off-loading is therefore included elsewhere in the pricing of the bill of quantities.

All timbers are assumed to be untreated whitewood, which is sawn, regularized or wrot as appropriate.

Example 11.1

Gang nailed trussed rafter, 35° pitch and 8 m span; timber assumed to be regularized.	Nr

Trussed rafter deivered to site	£38.00
Unload and stack: 0.1 hours at £5.30	0.53
	38.53
Waste 5%	1.93
Squad hoists and sets: 0.4 carpenter hours at £7.16	2.86
	43.32

Nails: 0.05 kg at £0.92/kg	0.05
Waste 10%	0.01
	43.38
Profit and oncost 20%	8.68
Rate per truss	£52.06

Example 11.2

Whitewood, sawn sleeper joists, laid on plates.			m	

	150 × 38 mm	100 × 50 mm	125 × 50 mm	150 × 50 mm
Material cost per 100 m	£127.00	£102.00	£127.00	£150.00
Nails: 1 kg at £0.92	0.92	0.92	0.92	0.92
	127.92	102.92	127.92	150.92
Waste 10%	12.79	10.29	12.79	15.09
	140.71	113.21	140.71	166.01
Squad laying varying quantities per carpenter hour at £7.16:				
150 × 38 mm: 24 m	29.83			
100 × 50 mm: 24 m		29.83		
125 × 50 mm: 19 m			37.68	
150 × 50 mm: 16 m				44.75
	170.54	143.04	178.39	210.76
Profit and oncost 20%	34.11	28.61	35.68	42.15
	204.65	171.65	214.07	252.91
Rate per metre	£2.05	£1.72	£2.14	£2.53

Example 11.3

Whitewood, sawn upper floor joists, laid without plates.			m	

	150 × 38 mm	150 × 50 mm	175 × 50 mm	200 × 50 mm
Material cost per 100 m	£127.00	£150.00	£182.00	£216.00
Nails: 1 kg at £0.92	0.92	0.92	0.92	0.92
	127.92	150.92	182.92	216.92
Waste 10%	12.79	15.09	18.29	21.69
	140.71	166.01	201.21	238.61

Squad laying varying quantities
    per carpenter hour at £7.16:

150 × 38 mm: 10.50 m	68.19		
150 × 50 mm: 10.00 m		71.60	
175 × 50 mm: 9.15 m			78.25
200 × 50 mm: 8.00 m			

	208.90	237.61	279.46	328.11
Profit and oncost 20%	41.78	47.52	55.89	65.62
	250.68	285.13	335.35	393.73
Rate per metre	£2.51	£2.85	£3.35	£3.94

(200 × 50 mm column: 89.50)

Example 11.4

> 100 × 38 mm redwood sawn Tanalized wall plate handed to builder for bedding.   m

Material cost per 100 m	£96.00
Labour handling: 0.3 hours at £5.30	1.59
	97.59
Profit and oncost 20%	19.52
	117.11
Rate per metre	£1.17

Example 11.5

> Whitewood: sawn ceiling joists and flat roof joists, laid on plates; and rafters/collars/struts, in stick built roofs.   m

	150 × 38 mm	100 × 50 mm	125 × 50 mm	150 × 50 mm
Material cost per 100 m	£127.00	£102.00	£127.00	£150.00
Nails: 1.8 kg at £0.92	1.66	1.66	1.66	1.66
	128.66	103.66	128.66	151.66
Waste 10%	12.87	10.37	12.87	15.17
	141.53	114.03	141.53	166.83

Squad laying varying quantities
    per carpenter hour at £7.16:

	150 × 38 mm	100 × 50 mm	125 × 50 mm	150 × 50 mm
150 × 38 mm: 16 m	44.75			
100 × 50 mm: 16 m		44.75		
125 × 50 mm: 13 m			55.08	
150 × 50 mm: 11 m				65.09
	186.28	158.78	196.61	231.92
Profit and oncost 20%	37.26	31.76	39.32	46.38
	223.54	190.54	235.93	278.30

Rate per metre	£2.24	£1.91	£2.36	£2.78

Example 11.6

Whitewood, sawn purlin/truss member			m	

	150 × 75 mm	100 × 100 mm	200 × 100 mm	300 × 100 mm
Material cost per 100 m	£240.00	£238.00	£460.00	£670.00
Nails: 1.8 kg at £0.92	1.66			
2.0 kg		1.84		
2.3 kg			2.12	
2.5 kg				2.30
	241.66	239.84	462.12	672.30
Waste 10%	24.17	23.98	46.21	67.23
	265.83	263.82	508.33	739.53
Squad laying varying quantities per carpenter hour at £7.16:				
150 × 75 mm: 9.5 m	75.37			
100 × 100 mm: 5.0 m		143.20		
200 × 100 mm: 2.5 m			286.40	
300 × 100 mm: 1.7 m				421.18
	341.20	407.02	794.73	1160.71
Profit and oncost 20%	68.24	81.40	158.95	232.14
	409.44	488.42	953.68	1392.85
Rate per metre	£4.09	£4.88	£9.54	£13.93

Example 11.7

Whitewood, sawn solid strutting to upper floor or flat roof joists.		m	

Joist depth	150 mm	175 mm	200 mm
Material 100 m			
125 × 50 mm at £127	£127.00		
150 × 50 mm at £150		£150.00	
175 × 50 mm at £182			£182.00
Nails: 2.80 kg at £0.92/kg	2.58	2.58	2.58
	129.58	152.58	184.58
Waste 10%	12.96	15.26	18.46
	142.54	167.84	203.04
Labour: 12.5 hours per 100 m at £7.16	89.50	89.50	89.50
	232.04	257.34	292.54
Profit and oncost 20%	46.41	51.47	58.51
	278.45	308.81	351.05
Rate per metre	£2.78	£3.09	£3.51

Example 11.8

Whitewood, sawn Herringbone strutting to upper floor or flat roof joists with 38 × 32 mm sawn whitewood; assume joists at 450 mm centres.	m

Joists at 450 mm centres have 400 mm spaces. There are 100/0.45 = 222 spaces. A 150 mm joist requires 2 struts approximately 450 mm long; total strutting is 222 × 2 × 0.45 ≃ 200 m. For 175 mm joists use 209 m, and for 200 mm joists use 218 m.

Joist depth	150 mm	175 mm	200 mm
Material:			
200 m at £22 per 100 m	£44.00		
209 m at £22 per 100 m		£45.98	
218 m at £22 per 100 m			£47.96
Nails: 4.3 kg at £0.92	3.96	3.96	3.96
	47.96	49.94	51.92
Waste 10%	4.80	4.99	5.19
	52.76	54.93	57.11
Labour: 20 hours per 100 m of two struts at £7.16	143.20	143.20	143.20
	195.96	198.13	200.31
Profit and oncost 20%	39.19	39.63	40.06
	235.15	237.76	240.37
Rate per metre	£2.35	£2.38	2.40

Example 11.9

Galvanized pressed steel herringbone strutting; assume joists at 450 mm centres.	m

Steel strutting fits most common centres and joist depths.
There are 222 spaces in 100 m, 2 struts per space and 2 nails per strut.

Material: 444 struts at £0.50 each	£222.00
Nails, 38 mm galvanized clout: 888 at 515/kg at £1.80/kg	3.10
	225.10
Waste 5%	11.26
	236.36
Labour: 12 m/hour at £7.16	59.67
	296.03
Profit and oncost 20%	59.21
	355.24
Rate per metre	£3.55

Example 11.10

145 × 15 mm redwood wrot fascia board and barge board grooved on back for soffit board.	m

Material cost per 100 m	£180.00
Nails: 0.5 kg at £1.32/kg	0.66
	180.66
Waste 10%	18.07
	198.73
Labour fixing 22 m per hour at £7.16	32.55
	231.28
Profit and oncost 20%	46.26
	277.54
Rate per metre	£2.78

Example 11.11

32 × 20 mm sawn Tanalized timber plugged and screwed to masonry.	m

Material cost per 100 m	£17.94
Screws at 100/1.2 m centres: 83 at £6.50 per 200	2.71
Plugs: 83 at £2.30 per 100	1.91
	22.56
Waste 10%	2.26
	24.82
Labour fixing 13 m per hour at £7.16	55.08
	79.90
Profit and oncost 20%	15.98
	95.88
Rate per metre	£0.96

Example 11.12

8 mm thick Douglas fir WBP plywood in soffit plate, let into groove in fascia/barge board and pinned to bearer; in width not exceeding 300 mm.	m

From the drawing the estimator sees that the width required is 165 mm. Sheets are supplied 2400 × 1200 mm. There are 1200/165 = 7 strips; 2400 × 45 mm is wasted by the saw or left over. This wastage of 3.75 per cent is included in the following material cost.

Material cost for 16.8 m strip from 1 sheet	£6.00
Nails: 0.2 kg at £2.30/kg	0.46
	6.46
Waste 10%	0.65
	7.11
Labour fixing 15 m per hour at £7.16	8.02
Cost for 16.8 m	15.13
Profit and oncost 20%	3.03
	18.16
Rate per metre	£1.08

An important departure from previous methods of measurement is the requirement to measure linings by the metre length of wall lined. In stud partition work the partition is measured by the metre length, stating the height (given in stages of 300 mm). The price must include the runners, studs and dwangs plus the plasterboard or other lining on both sides.

The calculation of the number of wall straps, studs or timbers fixed at centres presents a problem. The reason for requiring a fixing centre is that the sheet material which is being fastened to these timbers is of fixed dimension, e.g. gypsum wallboard in 1200 mm wide sheets is most conveniently fixed to timbers at 400 mm centres, i.e. joints between adjacent sheets fall on a timber. How long the wall or wide the ceiling happens to be determines the number of timbers necessary to provide this support.

For example:

An area 8 m wide requires $(8000/400)+1 = 21 = 2.625$ studs per metre
An area 6 m wide requires $(6000/400)+1 = 16 = 2.667$ studs per metre
An area 4 m wide requires $(4000/400)+1 = 11 = 2.750$ studs per metre
An area 3 m wide requires $(3000/400)+1 = 9 = 3.000$ studs per metre*
An area 2 m wide requires $(2000/400)+1 = 6 = 3.000$ studs per metre
An area 1 m wide requires $(1000/400)+1 = 4 = 4.000$ studs per metre*

* In both these instances the dividend is rounded up – no one fixes 'half' a batten.

So it would help to know the approximate length of individual partitions etc. However, it is impractical to calculate the number of studs or battens for every one so we add a percentage based on an appraisal of the mix of lengths of partitions etc. In the examples here we have added 10 per cent. So for a metre length of partition with studs at 400 mm centres there are exactly 2.5 studs plus 10 per cent = 2.75 studs × height of the partition.

Example 11.13

> Stud partition in height not exceeding 2.40 m, comprising 75 × 38 mm whitewood sawn top and bottom runners, studs at 400 centres and three rows of dwangs; 9.5 mm gypsum wallboard both sides. Jointing and decoration by others.   m

Pricing of this item is less prone to error if the logical building sequence is followed.

75 × 38 mm sawn softwood in runners

Material cost per 100 m	£69.00	
Nails: 0.8 kg at £0.92	0.74	
	69.74	
Waste 10%	6.97	
	76.71	
Squad fixing 20 m per hour at £7.16	35.80	
Cost of 100 m runner	112.51	
Cost for 2 m (top and bottom runners)		£2.25

75 × 38 mm sawn softwood in studding

Material cost per 100 m	69.00	
Nails: 1 kg at £0.92	0.92	
	69.92	
Waste 10%	6.99	
	76.91	
Squad fixing 16 m per hour at £7.16	44.75	
Cost of 100 m studding	121.66	

In a 1 m length of partition with studs at 400 centres there will be 2.5 m studs + 10% = 2.75.

Assume partition full 2.4 m high then there will be (2.4 × 2.75) = 6.6 m of studding

Cost of 6.6 m studding		£8.03

75 × 38 mm sawn softwood in dwangs (three in height)

Material cost per 100 m	69.00	
Nails: 1.4 kg at £0.92	1.29	
	70.29	
Waste 10%	7.03	
	77.32	
Squad fixing 8 m per hour at £7.16	89.50	
Cost of 100 m dwangs	166.82	

In a 1 m length of partition there will be 3 m of dwangs.

Cost of 3 m of dwangs		£5.00

9.5 mm thick gypsum wallboard nailed to softwood

Material cost per 100 m^2	117.00	
Nails: 3.75 kg at £2.20	8.25	
	125.25	
Waste 10%	12.52	
	137.77	
Squad fixing 6.5 m^2 per hour at £7.16	110.15	
Cost of 100 m^2 plasterboard	£247.93	

Unit length of partition 2.4 m high requires 4.8 m²

Cost of 4.8 m² of wallboard	£11.90
	27.18
Profit and oncost 20%	5.44
Rate per metre	£32.62

Note that in this example waste is assumed to be a constant 10%. If different percentages were required for the individual components then waste would have to be taken on the cost of the individual materials throughout the calculation.

If the carpenter has no lining material to fix (i.e. where it is fixed by another, such as the plasterer), measurement of battens and open spaced grounds may be expected to follow the traditional route and be measured by the square metre. The following examples show how this work might be priced.

Example 11.14

38 × 19 mm softwood vertical battens at 400 mm centres, plugged to masonry.   m²

Note that the plugging referred to here is achieved by cutting short lengths of timber and driving these into pockets cut into mortar joints. To cover 1 square metre with battens at 400 mm centres requires (1000/400)=2.5 m material plus ends.

Material cost per 100 m	£16.50
Nails: 0.3 kg at £0.92	0.28
Timber for plugs 75 mm long, at 0.8 m centres = 125 plugs: allowing for ends, say 140 × 75 mm = 10.5 m	1.73
	18.51
Waste 10%	1.85
	20.36
Squad plugging and fixing 4 m per carpenter hour at £7.16	179.00
	199.36
Profit and oncost 20%	39.87
	239.23
Add for 'ends'—10%	23.92
	£263.15
Rate per m² (2.5 m of material)	£6.58

Example 11.15

38 × 19 mm sawn softwood branders to ceilings joists, fixed at 400 mm centres.	m²

Material cost per 100 m	£16.50
Nails: 0.3 kg at £0.92	0.28
	16.78
Waste 10%	1.68
	18.46
Squad fixing 11 m per carpenter hour at £7.16	65.09
	83.55
Profit and oncost 20%	16.71
	100.26
Add for 'ends'—10%	10.03
	110.29
Rate per m² (2.5 m of material)	£2.76

12

Woodwork joinery : second fixings and finishings

SMM7 Sections K11, K20, L10, L20, L30, P20, P21

This part of the estimating for woodwork in modern practice is now very much more concerned with the fixing of pre-finished units. The idea of these units is not new. However, the rapid development of new materials and technologies, together with economic pressures on the building industry, have accelerated the demand for and use of pre-finished units. The joiner today is no less skilful than his predecessor, but has new skills and techniques which have boosted his output.

Floors and decks are still boarded with softwood and hardwood strip. However, there are sheet materials now available which one can almost consider to be pre-finished units laid over joists. Indeed, some of these sheet materials have peelable coverings to facilitate leaving a clean floor or deck after the works have been completed.

Doors can now be obtained not just pre-finished but complete in their frames, with architraves fixed one side and loose the other. The frame is placed into the opening in the wall or partition and fastened with frame anchors or screws. The protective polythene wrapper is peeled off, and the door is there complete with all its ironmongery!

Similarly, windows are coming on site already glazed and with all ironmongery factory fitted. They are fixed into openings in walls with framing anchors or brackets and the protective wrapping is removed. The window is then fully functional during building, keeping the weather out and letting light in. Even window boards and bed moulds are part of the kit to be added after fixing the window into position.

The multitude of fasteners available have already been discussed in the previous chapter, as have the various difficulties encountered in working with softwoods and hardwoods.

Before going on to some examples of rates, the reader should remember that we have been working with a squad of 1 labourer and 5 craftsmen costing £7.16 per craftsman hour. This hourly rate will continue to be used in this chapter.

Example 12.1

20 mm thick particle board flooring, tongued and grooved all round and nailed to joists with 65 mm improved nails.	m^2

The cover size per sheet is 2400×600 mm.

Material cost per 100 m²	£412.00

Nr of sheets per 100 m²: $100/(2.4 \times 0.6) = 69.44$
Nails per sheet with joists at 400 mm centres: 28

Nails per 100 m²: 1944 at 229/kg at £27 per 25 kg	9.17
	421.17

Waste 5%	21.06
	442.23
Labour: craftsman fixes 5 m² per hour at £7.16: 100 m²	143.20
	585.43
Cost per m²	5.85
Profit and oncost 20%	1.17
Rate per m²	£7.02

Example 12.2

An alternative method of calculation for the item in Example 12.1 would be based on the cost per m² for the particle board sheets, as follows.

Material cost per m²	£4.12
Nails per m²: 20 at 229/kg at £27 per 25 kg	0.09
	4.21
Waste 5%	0.21
	4.42
Labour: craftsman fixes 5 m² per hour at £7.16: 1 m²	1.43
Cost per m²	5.85
Profit and oncost 20%	1.17
Rate per m²	£7.02

Example 12.3

19 mm thick softwood tongued and grooved flooring twice nailed to joists; joists at 450 mm centres.	m²

Material cost per 100 m, 150 mm wide	£115.80
Nails: 65 mm bullet headed brads, 2 per joist, 245* joists per 100 m: 490 at 229/kg at £24 per 25 kg	2.05
	117.85
Waste 5%	5.89
Material cost for 15 m²	123.74
Material cost for 1 m²	8.25
Labour: craftsman lays 4 m² per hour at £7.16: 1 m²	1.79
	10.04
Profit and oncost 20%	2.00
Rate per m²	£12.04

*Includes 10% for 'ends'.

Example 12.4

> 3.2 mm thick standard hardboard pinned to open spaced grounds with
> 15 mm zinc plated lost head panel pins.   m^2

Grounds are at 400 mm centres both ways and pins are spaced at 50 mm centres.

Material cost per m^2	£1.25
Nails: framing at 400 mm centres gives 2.75 m* of nailing each way at 50 mm centres:	
110 panel pins at 900/kg at £1.40/kg	0.17
	1.42
Waste 5%	0.07
	1.49
Labour: craftsman fixes 4 m^2 per hour at £7.16: 1 m^2	1.79
	3.28
Profit and oncost 20%	0.66
Rate per m^2	£3.94

* Includes 10% for 'ends'.

Example 12.5

> Standard casement window in pre-treated softwood to BS 644 Part 1,
> type 130V, overall size 600 mm wide × 900 mm high, fully weather
> stripped and fitted into opening in brick wall with proprietary framing
> fasteners.   Nr

Casement window	£29.72
Fischer fasteners: 4 at £0.43	1.72
	31.44
Waste 2.5%	0.79
	32.23
Labour: joiner fixes window in 0.5 hours at £7.16	3.58
	35.81
Profit and oncost 20%	7.16
Rate per window	£42.97

Example 12.6

Standard casement window in pre-treated softwood to BS 644 Part 1, type 240 TW, overall size 1200 mm wide × 1200 mm high, fully weather stripped and fitted into opening in brick wall with framing fasteners as Fischer's XYZ.	Nr

Casement window	£59.25
Fischer fasteners: 6 at £0.43	2.58
	61.83
Waste 2.5%	1.55
	63.38
Labour: joiner fixes window in 0.7 hours at £7.16	5.01
	68.39
Profit and oncost 20%	13.68
Rate per window	£82.07

Example 12.7

Satin anodized aluminium casement window stay and two pins, screwed to softwood with matching screws.	Nr

Window stay, pins and screws	£3.15
Waste 5%	0.16
	3.31
Labour: joiner fixes set in 0.25 hours at £7.16	1.79
	5.10
Profit and oncost 20%	1.02
Rate per set	£6.12

Example 12.8

Satin anodized aluminium casement window fastener, screwed to softwood with matching screws.	Nr

Window fastener and screws	£2.70
Waste 5%	0.13
	2.83
Labour: joiner fixes set in 0.35 hours at £7.16	2.51
	5.34
Profit and oncost 20%	1.07
Rate per set	£6.41

Example 12.9

> Prefabricated wrot rebated softwood door frame, ex 63 × 88 mm, for
> door size 838 × 1981 mm.   Nr

Frame	£23.00
Frame fasteners: 6 at £0.43	2.58
	25.58
Waste 2.5%	0.64
	26.22
Labour: joiner fixes frame into brick opening: 0.5 hours at £7.16	3.58
	29.80
Profit and oncost 20%	5.96
Rate per frame	£35.76

An alternative form of frame is made up from rectangular section timber and provided with stops measured under a separate item.

Example 12.10

> Wrot softwood door frame, ex 63 × 38 mm, for door size 838 × 1981 mm,
> plugged to brickwork.   m

Wrot softwood 63 × 38 mm

Posts	2 at	1981	3962	
Head	1 at	838	838	
Joints	2 at	25	50	
Legs	2 at	63	126	
Horns	2 at	50	100	
		5.076 m at £72 per 100 m		£3.65

Timber for plugs ex 63 × 38 mm, 6 at 75 mm: 0.45 m	0.32
Nails: if frame double nailed at three points each leg then 12 nails for fixing. 6 nails required for making.	
18 nails at 150/kg at £22 per 25 kg	0.11
	4.08
Waste 10%	0.41
	4.49
Labour: joiner makes up 2.5 m per hour at £7.16	14.54
Plugging to brickwork: 0.75 hours at £7.16	5.37
	24.40
Profit and oncost 20%	4.88
Cost per frame (compare Example 12.9)	29.28
Rate per metre	£5.77

Example 12.11

44 mm thick cellular core flush panelled external quality door, overall size 838 × 1981 mm, faced with plywood and with hardwood lippings on vertical edges, finished for painting, hung on 1.5 pairs 100 mm brass butt hinges.	Nr

Door	£49.60
Hinges: 1.5 pairs at £7.50 per pair	11.25
Screws, nr 8 brass 40 mm: 8 per hinge at £12 per 200	1.44
	62.29
Waste 2.5%	1.56
	63.85
Labour: joiner hangs door in 1.5 hours at £7.16	10.74
	74.59
Profit and oncost 20%	14.92
Rate per door	£89.51

Example 12.12

35 mm thick cellular core flush panelled external quality door overall size 838 × 1981 mm, faced with hardboard and with hardwood lippings on vertical edges, hung on 1 pair 100 mm brassed steel butt hinges.	Nr

Door	£29.60
Hinges: 1 pair at £0.90 per pair	0.90
Screws, nr 8 brassed 40 mm: 8 per hinge at £4 per 200	0.32
	30.82
Waste 2.5%	0.77
	31.59
Labour: joiner hangs door in 1.25 hours at £7.16	8.95
	40.54
Profit and oncost 20%	8.11
Rate per door	£48.65

In Examples 12.13 and 12.14, the cost of the ironmongery item fixed is included in the rate. It is more usual to have the supply of the ironmongery the subject of a prime cost sum, and to give fix-only items in the bill of quantities. This is illustrated in Examples 12.15 and 12.16.

Example 12.13

5 lever mortice lock and satin anodized aluminium lever furniture and escutcheons fixed with matching screws.	Nr

Lock	£16.00
Lever furniture	13.00
Escutcheon plates and screws	4.00
	33.00
Waste 5%	1.65
	34.65
Labour: joiner fixes complete in 1.5 hours at £7.16	10.74
	45.39
Profit and oncost 20%	9.08
Rate per set	£54.47

Example 12.14

Satin anodized aluminium postal plate including slotting door.	Nr

Postal plate and fixings	£7.00
Waste 5%	0.35
	7.35
Labour: joiner fixes complete in 1.2 hours at £7.16	8.59
	15.94
Profit and oncost 20%	3.19
Rate per plate	£19.13

Example 12.15

Fit and fix only mortice lock, lever furniture and escutcheon plates.	Nr

Material cost included in prime cost sum	nil
Labour: joiner fixes in 1.5 hours at £7.16	£10.74
Profit and oncost 20%	2.15
Rate per set	£12.89

Example 12.16

Fit and fix only hat and coat hook.	Nr

Material cost included in prime cost sum	nil
Labour: joiner fixes in 0.08 hours at £7.16	£0.57
Profit and oncost 20%	0.11
Rate per hook	£0.68

Example 12.17

> Open tread staircase 914 mm wide, in Parana pine complying with
> BS 585:Part 1, BS 5395; total going 2700 mm; with 237 × 32 mm
> strings, 258 × 32 mm treads, 110 × 20 mm dummy risers,
> 66 × 66 × 1350 mm newel posts top and bottom, newel caps,
> 63 × 44 mm handrail 32 × 32 mm balusters at 95 mm centres; one string
> plugged to wall, other string housed both ends to newels, newels
> checked to joists; the whole left clear for varnishing.   Nr

Materials

Stair		£440.00
Newel posts:	2 at £17.50	35.00
Newel caps:	2 at £4.80	9.60
String capping:	2 at £12.00	24.00
Handrail:	1 at £14.00	14.00
Balusters:	29 at £1.40	40.60
Plugs: 8 nr 75 mm long ex 100 × 38 mm at £77 per 100 m		0.46
Glue: say		1.00
Nails 50 mm oval brads: say		0.30
		564.96
Waste 2.5%		14.12
		579.08

Labour

Joiner to fit and fix prefabricated stair: 3.5 hours at £7.16	25.06
Fix newels: 2 at 1.35 m at 2 m per hour at £7.16	9.67
Fix capping to string: 1 at 3.9 m at 4 m per hour at £7.16	6.98
Fix balusters: 29 at 0.15 hour each at £7.16	31.15
Fix newel cappings: 2 at 0.05 hour each at £7.16	0.72
	652.66
Profit and oncost 20%	130.53
Rate per staircase	£783.19

Example 12.18

> Skirting 100 × 19 mm with two labours, nailed to softwood.   m

Materials per 100 m	£67.00
Nails: 50 mm oval brads, 2 at 400 mm centres:	
550* per 100 m, at 440/kg at £33 per 25 kg	1.65
	68.65
Waste 5%	3.43
	72.08
Labour: joiner fixes 6.25 m per hour at £7.16	114.56
	186.64
Profit and oncost 20%	37.33
	223.97

Rate per metre   £2.24

*Includes 10% for 'ends'.

Plugging to masonry can be calculated in the same fashion as earlier examples, whether using proprietary plugs or timber plugs.

Example 12.19

Architrave 100 × 19 mm with two labours, nailed to softwood.	m

Materials per 100 m	£67.00
Nails: 50 mm oval brads, 2 at 400 mm centres:	
550 per 100 m, at 440/kg at £33 per 25 kg	1.65
	68.65
Waste 5%	3.43
	72.08
Labour: joiner fixes 8 m per hour at £7.16	89.50
	161.58
Profit and oncost 20%	32.32
	193.90
Rate per metre	£1.94

Example 12.20

Door stop 42 × 19 mm, nailed to softwood.	m

Materials per 100 m	£37.00
Nails: 40 mm oval brads, 2 at 400 mm centres:	
550 per 100 m, at 440/kg at £44 per 25 kg	2.20
	39.20
Waste 5%	1.96
	41.16
Labour: joiner fixes 10 m per hour at £7.16	71.60
	112.76
Profit and oncost 20%	22.55
	135.31
Rate per metre	£1.35

13

Structural steelwork and metalwork

SMM7 Sections G10, G12, L11, L21, L31

Introduction

Structural steelwork and metalwork are specialist elements of the building and as such are normally the subject of a subcontract. It would be extremely rare for the contractor's estimator to be involved in pricing other than, for example, an isolated steel section to be used as a lintel, or the fixing in position of a fabricated metal window or metal balustrade.

SMM7 takes account of the specialist nature of these elements of construction in its requirements for dimensioned drawings from which the specialist can measure the work involved. The relevant SMM7 sections are as follows:

G10 Structural steel framing
G12 Isolated structural metal members
L11 Metal windows/rooflights/screens/louvres
L21 Metal doors/shutters/hatches
L31 Metal stairs/walkways/balustrades.

This chapter deals with a selection of these elements.

Isolated structural metal members (G12)

Isolated structural metal members are the most common example of steelwork fixed in position by the contractor's own operatives. Proprietary steel lintels over window and door openings will be fixed in position by bricklayers in the course of their work. Other heavier sections may require craneage.

Example 13.1

Isolated structural plain member in universal beam size 356×171 mm $\times 67$ kg/m, as lintel over 4 m door opening, and bolting to padstones (measured separately).   tonnes

Labour/hire
For hoisting the beam into position allow:

Labourer 1 hour at £5.30	£5.30
Bricklayer 1 hour at £6.10	6.10
Hire 750 kg mobile crane and operative: 4.5 hours at £12.50	56.25
Fixing bolts: 8 at 0.1 hours each at £6.10	4.88
	72.53

Expressed as a rate per tonne: £72.53/0.31   233.97

Materials
Steel beam cost delivered to site: £350 per tonne   350.00
Fixing bolts included with brickwork   nil
                                                                           ———
                                                                           583.97
Profit and oncost 20%   116.79
                                                                           ———
Rate per tonne   £700.76

Strictly, proprietary galvanized steel lintels should be measured by weight, however, clause 6.1 of the General Rules in SMM7 allows proprietary items to be measured by their catalogue reference, and this is assumed in the following example.

The squad is assumed to comprise 2 bricklayers and 1 labourer (see Chapter 7), with a cost of £17.50 per hour.

Example 13.2

Lintel type T1, 155 mm high × 2100 mm long.	Nr

Labour fixing at 0.18 hours/m: 0.378 hours at £17.50   £6.61
Material cost   33.50
                                                                           ———
                                                                           40.11
Profit and oncost 20%   8.02
                                                                           ———
Rate per lintel   £48.13

Metal windows/rooflights/screens/louvres (L11)

Metal windows are supplied to site complete and ready for fixing to a timber subframe. Aluminium windows are supplied fully finished and often pre-glazed. Example 13.3 assumes that a timber subframe is already in position and has been priced under the timber item L10.

Aluminium windows are delivered to site with tape protecting the finish and with labelling on the glass. The labelling is to inform operatives that the windows are glazed, and helps to prevent, for example, scaffolding tubes being passed through the glazed window.

Pre-glazed units will require careful handling, storing and placing in position, which is reflected in the rate.

Example 13.3

Type XYZ aluminium window, size 888 × 888 mm, with white acrylic finish, factory clear glazed with 11 mm double glazing units, fixed with screws to timber subframe (separately measured) and finished externally with mastic filler.	Nr

Labour

Labourer taking delivery, storing, placing in position: 0.75 hours at £5.30	£3.98
Joiner fixing: 1 hour at £6.10	6.10
Bricklayer mastic: 0.5 hours at £6.10	3.05
Labourer removing tape and cleaning glass: 0.75 hours at £5.30	3.98

Materials

Pre-glazed window	82.00
Damage risk 5%	4.10
Mastic and screws: say	2.00
	105.21
Profit and oncost 20%	21.04
Rate per window	£126.25

Metal stairs/walkways/balustrades (L31)

In a similar manner to the metal window described above, metal stairs and balustrades are generally measured by number and by reference to a drawing or a manufacturer's catalogue. If the items are supplied for fixing by the contractor, the estimator must liaise with the manufacturer to determine the method of fixing anticipated. It should be noted that mortices in concrete are measured under SMM7 Section E20 or E41.

It is normal for balustrades and the like to be in sections which are capable of being moved and placed in position by hand. However, if a particularly large component, such as a fire escape staircase, requires mechanical lifting, then the estimator should include for this in the item rate.

14

Plumbing installations

SMM7 Sections R10, R11, Y10, Y11, N13

Introduction

This chapter covers standard plumbing installations. Mechanical heating/cooling/refrigeration systems, and ventilation and air conditioning systems, are more specialized forms of mechanical engineering and are generally carried out by specially trained tradesmen. Examples for these systems will not be given. However, the principles of the plumbing examples can be applied to these other systems, bearing in mind the differences in wage costs and the need for specialized tools and plant, especially for handling bulky or heavy items.

The SMM7 sections are: rainwater pipework/gutters; foul water drainage above ground; pipelines; pipeline ancillaries; and sanitary appliances/equipment.

All-in rates for pipe

Pipe is available in a bewildering variety of materials and types. For the plumber, plastics, metal and composition materials are the most common. In these materials, pipes are manufactured to carry hot and cold water; other liquids, including those that are corrosive, flammable, toxic etc.; gases; foul water, rain water and waste water; and so on. Such pipes share certain features which make it possible to treat them as a single class for measurement, and also allow us to generalize in the explanation of how to determine all-in rates.

Pipe is supplied in lengths or coils – rigid or brittle materials being in lengths and flexible materials being in coils. The most common length is 6 m. However, some pipe is supplied in shorter or longer lengths, especially where the joint is formed as part of the pipe, as in PVC or cast iron soil pipe with spigot and socket joints. For these types, variety of lengths (say 1, 2, 3 and 4 m) reduces waste, because every cut produces a length without a joint!

Coils of 25, 50, 100 or 150 m are common, but some pipe is supplied in 30 m or even 60 m coils. The coil length is dictated as much by what the plumber wants to keep in stock as by old traditional manufacturing processes. For example, it might not have been possible to extrude more than 30 m of a particular pipe in one operation, and therefore it became standard practice to supply in coils of that length. The pipe must inevitably be cut to length. The shorter are the cut lengths, the greater is the labour, the fewer are the joints in the running length and, depending on the rigidity of the pipe and the appliances served, the fewer are the fastenings to hold the pipe in place.

Pipe material varies as follows in its bending ability:

- Bends can be easily made, e.g. in polyethylene water pipe (although of large radius).
- Bends can be made with apparatus, e.g. in half hard copper pipe (BS 2871 Table X) for hot and cold water supply, using springs or a light bending machine.
- Bends cannot be made and elbows or bent couplings have to be used, e.g. in cast iron pipe, uPVC soil pipe or hard thin wall copper pipe (BS 2871 Table Z).

The material is always fully specified so that identification is positive.

There is generally a standard method of fixing the pipe in position. Non-standard fixings must be fully specified in the bill item or preamble. The background to which the pipe is fixed is also given. Backgrounds are tabulated in the general rules to SMM7, 8.3(a) to (e).

Joints

Although joints in the running length are deemed to be included, there is a standard method of jointing for each type of pipe in each situation.

Joints occur of course at fittings, i.e. junctions, bends, elbows, valves etc., and where pipe is connected to sanitary fittings, tankage etc. In every instance there is a standard method of forming these joints and connections and they are measured separately from the pipe, although some may be included with the apparatus.

Here it may be as well to define the two types of joint that a plumber can make. A coupling is designed to join pipe to pipe only. A connection is used to join pipe to sanitary fittings, tanks, cylinders etc. The term 'joint' will be used in this text to mean either type.

When making joints the tradesman will use very small (per joint) amounts of general materials. For example, capillary joints on copper pipework require steel wool to clean the pipe, flux to assist in making the solder flow into the whole joint, and fuel to provide heat for the blow torch or clamps. Capillary joints may or may not have an integral solder ring, and if not then solder must be allowed for. Compression joints will require either jointing compound or PTFE tape to ensure a water/gas tight joint. The ring sealed push fit joints on PVC waste pipe require a lubricant. The threaded joints in gavanized steel pipework require hemp yarn and jointing compound or PTFE tape. Joints in spigot and socketed cast iron pipe require gaskin, molten lead and fuel for the furnace. The list goes on and on.

In practically every instance very small amounts of material are used which would be tedious to quantify for every joint. The usual practice is to make a monetary allowance for items like PTFE tape, jointing compound etc., either in the preliminary works bill or as a small percentage addition at the time of building up the rate. Easily quantifiable and more expensive materials such as solder in end feed joints, or lead and yarn for cast iron soil pipes, are actually quantified and priced when building up the rate for joints of that kind.

SMM7 has recognized that joints are important to the estimator by classifying couplings and connections as having one, two three or more ends. However, it has made costing difficult for the estimator by lumping together all fittings up to and including 65 mm diameter in one item. (See, for example, rule 2.3.1 in Section R on disposal systems.) The difference in cost between a brass/gunmetal 15 mm tee and one of 28 mm is quite significant. More of this is illustrated in the examples which follow.

Special tools

The plumber requires many special tools besides spanners, grips etc.: for example, blow torches (now generally fuelled with butane or propane), pipe bending machines, pipe threading machines and so on. It would not be sensible to attempt to allocate the costs for these tools over each individual item in the bill of quantities. The best solution is to put the cost into the preliminaries bill. Fuel in connection with any of these tools or small plant items might be included here rather than in the rates. The basis for the calculation of the amount of fuel is the general size of the contract and knowledge of fuel consumption on previous contracts.

Labour costs and outputs

In Chapter 2 it was noted that plumbers are paid at a higher rate than other operatives. Thus all-in hourly rates have to be established for this section of work for the various grades of skilled, semi-skilled and unskilled workers. Plumbers are generally classified as technical, advanced and trained; the first two grades carry out various degrees of supervision. The plumber generally operates with a mate, who may be a lower or similar grade of plumber, or even an unskilled worker or an apprentice. Wages for plumbers, apprentices and mates are generally agreed by the Joint Industry Board for the Plumbing and Mechanical Engineering Services Industry.

The first thing to consider is whether or not the labour constants used are for one craft operative working on his own (this is rare), for one craft operative working in a gang, or for a gang. If the constants are for a gang then the composition of the gang must be known, e.g. two plumbers, a plumber and an apprentice, two plumbers and a labourer, and so on.

Next the cost of this labour has to be considered. This is best illustrated with examples. To set the scene it will be assumed that all-in hourly rates for the following operatives have been calculated:

Plumbers:	technical	£8.90
	advanced	£7.85
	trained	£7.45
Apprentices:	third year	£4.35

Now, a single plumber is costed out per hour at his particular rate. For two plumbers in a gang, the gang is costed out as the sum of the individual rates. However, if the constants are for one plumber then the average rate of the individual rates is used. For example:

	Gang rate	Individual rate
1 plumber	—	As grade
2 plumbers: 1 advanced, 1 trained	£15.30	£7.65
2 plumbers (1 advanced, 1 trained) and third year apprentice	£19.65	£9.83

It will be noticed that in the calculation of the third individual rate, the contribution made by the apprentice to the gang's work is not included; the gang rate is divided by the number of qualified plumbers. This practice of ignoring the apprentice's contribution is common among estimators. Some do include such a contribution; however, it cannot be equivalent to a full craft operative. First of all, apprentices generally do not work as long a day as craft operatives. A third year plumbing apprentice only works 80 per cent of the hours of a plumber. Secondly, apprentices are allowed time off to attend technical college. Thirdly, they cannot be as productive simply because they are still learning the craft and require guidance. Finally, their training is given by the craft operative(s) and this will slow the operative down. On balance, if apprentices are included in a squad it is probably best to ignore their contribution to production.

Rates used in the examples in this chapter are based on a gang of one advanced and one trained plumber, costing £15.30 per hour. All constants given in the tables and elsewhere are based on output per craft operative working in a gang.

Table 14.1 *Copper pipe to BS 2871, Table X, clipped to background including joints in the running length: range of outputs for plumber in a squad in hours per metre of pipe or per joint or labour*

	Pipe diameter		
	15 mm	22 mm	28 mm
Pipe:			
Screwed to timber background	0.19–0.29	0.20–0.30	0.20–0.30
Plugged and screwed to background	0.21–0.40	0.22–0.50	0.24–0.60
Made bends	0.10–0.20	0.15–0.33	0.24–0.60
Capillary joint:			
Elbow/bent coupling	0.18–0.20	0.18–0.32	0.24–0.40
Tee	0.23–0.26	0.23–0.44	0.30–0.40
Straight connection	0.22	0.22	0.28
Straight connection + back nut	0.28	0.28	0.33
Valve	0.20–0.60	0.20–0.70	0.28–0.85
Compression joint:			
Elbow/bent coupling	0.15–0.24	0.16–0.24	0.18–0.32
Tee	0.20–0.33	0.20–0.33	0.30–0.39
Straight connection	0.16	0.16	0.22
Straight connection + back nut	0.20	0.20	0.26
Valve	0.20–0.60	0.22–0.70	0.28–0.85

Copper pipework and fittings

Table 14.1 shows ranges of outputs for copper pipework and fittings. The outputs have been abstracted from a variety of sources and reveal the diversity of opinion common among estimators. There has to be an explanation for these differences. The variable attributable to labour factors have been well rehearsed elsewhere in the text. This section deals with those factors attributable to trade practice and to specification.

Variations in pipework installations

Take for example a pipe run along a wall in a warehouse or factory There will be an uninterrupted flow of work. The operatives will snap a chalk line on the background; fasten clips to the background and mount the pipe in the clips; make joints in the running length as they proceed; and finally tighten the clips. The question of what type of joint is made will be discussed later. The only cutting will be at the ends. The plumber can reasonably be expected to erect better than 5 m per hour in these circumstances.

By contrast, take the situation where the pipe is being erected in the toilet blocks in the same warehouse or factory. The first lengths erected during the plumber's roughing stage will present slightly more difficulty due to the restrictions on space. Further time will be required to set out tees to sanitary fittings. The pipe connecting the plumber's roughing to the sanitary fittings will be even more difficult. There will be many made bends which, although measurable and priced separately, will require the plumber to take more time cutting pipe to the correct length. There will also be proportionatey more cutting to length than in any other situation. Moreover, there will be work in confined spaces and under and around the sanitary fittings.

Fittings

BS 864 covers capillary and compression fittings suitable for the copper pipe under discussion. Two kinds of capillary fitting are manufactured, as well as two kinds of compression fitting. With both kinds of fitting there are many variations.

The first type of capillary fitting has a solder ring incorporated by the manufacturer. If a flux is used which cleans the pipe, fittings can be pushed onto the pipe and heat applied for a fast joint. It is hard to believe that it takes 12 minutes to form such a joint between two straight lengths of pipe. However, it must be remembered that the plumber is expected to carry out such tasks for 8 hours a day, 5 days a week – and to plan ahead for the next joint!

The second kind of capillary fitting has no solder in it. The pipe is smeared with flux (which again can be a self-cleaning flux), the fitting is pushed onto the pipe and heat is applied. A stick of solder must then be applied to each end of the fitting and allowed to melt and to be drawn into the joint. This is a much more skilful operation; it is easy to put in too little solder and have a joint that is leaky or weak under pressure, or to put in too much solder and allow it to flow into the pipe and cause an obstruction. The latter is obviously expensive in its use of solder – and both are expensive when they cause problems for the contractor which have to be put right! In this instance, the time of 12–15 minutes per joint seems quite fair.

The two types of compression joint are described as type A non-manipulative and type B manipulative. Manipulative means that the ends of the pipes to which the fitting is attached are worked to a special shape which makes the joint mechanically sound and pressure tight. Non-manipulative means that the fitting has been designed to be mechanically sound and pressure tight without the need to do anything to the pipe.

Example 14.1

> 15 mm copper pipe, BS 2871:Part 1, Table X, fixed at 1200 mm centres with copper pipe bands screwed to timber, including capillary couplings in the running length.   m

Pipe is generally bought in 6 m lengths; however, the basis for pricing is 100 m.

Materials

Pipe 100 m	£113.00
Clips: 100/1.2, say 84 at £1.95 per 100	1.64
Screws brass nr 8 × 25 mm: (100/1.2) × 2, say 168 at £4 per 100	6.72
Straight capillary couplings, integral solder ring:	
(100/6) + 1, say 18 at 12p each	2.16
	123.52
Waste 2.5%	3.09
	126.61

Labour

Pipe plumber fixes 5 m per hour: 100/5 at £7.65	153.00
Couplings: (100/6) + 1, say 18 at 0.22 hours at £7.65	30.29
	309.90
Profit and oncost 20%	61.98
	371.88
Rate per metre	£3.72

Example 14.2

> 22 mm copper pipe, BS 2871:Part 1, Table X, fixed at 1200 mm centres with copper pipe bands screwed to timber, including capillary couplings in the running length. m

Materials

Pipe 100 m	£130.00
Clips: 100/1.2, say 84 at £2.60 per 100	2.18
Screws brass nr 8 × 25 mm: (100/1.2) × 2, say 186 at £4 per 100	6.72
Straight capillary couplings, integral solder ring:	
(100/6) + 1, say 18 at 20p each	3.60
	142.50
Waste 2.5%	3.56
	146.06

Labour

Pipe plumber fixes 4 m per hour: 100/4 at £7.65	191.25
Couplings: (100/6) + 1, say 18 at 0.22 hours at £7.65	30.29
	367.60
Profit and oncost 20%	73.52
	441.12
Rate per metre	£4.41

Example 14.3

> 28 mm copper pipe, BS 2871:Part 1, Table X, fixed at 1400 mm centres with copper pipe bands plugged and screwed to masonry, including capillary couplings in the running length. m

Materials

Pipe 100 m	£180.00
Clips: 100/1.4, say 72 at £3.90 per 100	2.88
Screws brass nr 8 × 25 mm: (100/1.4) × 2, say 144 at £4 per 100	5.76
Plugs: (100/1.4) × 2, say 144 at £2.30 per 100	3.31
Straight capillary couplings, integral solder ring:	
(100/6) + 1, say 18 at 33p each	5.94
	197.82
Waste 2.5%	4.95
	202.77

Labour

Pipe plumber fixes 3 m per hour: 100/3 at £7.65	255.00
Couplings: (100/6) + 1, say 18 at 0.28 hours at £7.65	38.56
	496.33
Profit and oncost 20%	99.27
	595.60
Rate per metre	£5.96

Example 14.4

Joints, valves and small labours on copper pipework

Where pipe size ≤65 mm, SMM7 requires all sizes of such joints and labours to be included in one item. It follows that the rates given in this example are not bill rates. How the individual rates in the example are aggregated or proportioned to give a bill rate is explained after the example rates.

Extra over given diameter of copper pipe for made bend:

Pipe diameter	15 mm	22 mm	28 mm
Labour: 0.10 hours at £7.65	£0.77		
0.15 hours at £7.65		£1.15	
0.28 hours at £7.65			£2.14
Profit and oncost 20%	0.15	0.23	0.43
Each	£0.92	£1.38	£2.57

Extra over copper piping ≤65 mm diameter for capillary joints with two ends:

Pipe diameter	15 mm	22 mm	28 mm
Material: bent coupling	£0.20	£0.40	£0.70
Waste 2.5%	0.01	0.01	0.02
Labour: 0.18 hours at £7.65	£1.38		
0.22 hours at £7.65		£1.68	
0.30 hours at £7.65			£2.29
	1.59	2.09	3.01
Profit and oncost 20%	0.32	0.42	0.60
Each	£1.91	£2.51	£3.61

Extra over copper piping ≤65 mm diameter for capillary joints with three ends:

Pipe diameter	15 mm	22 mm	28 mm
Material: tee	£0.35	£0.70	£1.35
Waste 2.5%	0.01	0.02	0.03
Labour: 0.25 hours at £7.65	1.91		
0.30 hours at £7.65		2.30	
0.35 hours at £7.65			2.68
	2.27	3.02	4.06
Profit and oncost 20%	0.45	0.60	0.81
Each	£2.72	£3.62	£4.87

Straight connection with backnut, compression ends, type A, for copper:

Pipe diameter	15 mm	22 mm	28 mm
Material: connection	£0.60	£1.08	£1.80
Waste 2.5%	0.02	0.03	0.05
Labour: 0.35 hours at £7.65	2.68		
0.59 hours at £7.65		4.51	
1.35 hours at £7.65			10.33
	3.30	5.62	12.18
Profit and oncost 20%	0.66	1.12	2.44
Each	£3.96	£6.74	£14.62

Screwdown stop valve, compression joints, type A, for copper pipe:

Pipe diameter	15 mm	22 mm	28 mm
Material: valve	£1.38	£2.41	£6.29
Waste 2.5%	0.03	0.06	0.16
Labour: 0.30 hours at £7.65	2.29		
0.35 hours at £7.65		2.68	
0.42 hours at £7.65			3.21
	3.70	5.15	9.66
Profit and oncost 20%	0.74	1.03	1.93
Each	£4.44	£6.18	£11.59

Gate valve, compression joints, type A, for copper pipe:

Pipe diameter	15 mm	22 mm	28 mm
Material: valve	£7.54	£9.95	£12.98
Waste 2.5%	0.19	0.25	0.32
Labour: 0.30 hours at £7.65	2.29		
0.35 hours at £7.65		2.68	
0.42 hours at £7.65			3.21
	10.02	12.88	16.51
Profit and oncost 20%	2.00	2.58	3.30
Each	£12.02	£15.46	£19.81

Where pipe fittings are measured extra over the pipe in which they occur, the cost of the equivalent length of pipe should be deducted. Where these equivalent lengths are very small or the cost is relatively low, this deduction becomes academic. Accordingly, no deduction for this cost has been made in the rates given in Example 14.4.

Note also that when costing out the extra over items for fittings ≤65 mm diameter with two or three ends, there is a significant range of cost from 15 to 28 mm – and 65 mm fittings are much more expensive! The estimator will have to make allowance for this by assessing the ratio of fittings with one, two and three ends in each contract and proportioning out his rates accordingly.

For example, suppose a bill has 100 m of 15 mm pipework and 20 m of 22 mm pipework. There is one bathroom, one shower room and one WC compartment, plus a kitchen and a utility room. The odds are that most of the fittings with one, two and three ends are on 15 mm pipework, probably in the ratio of at least 5:1. Rates from Example 14.4 would be adjusted as follows:

Fittings with two ends:
    15 mm: 5 at £1.91   £9.55
    22 mm: 1 at £2.51   2.51

              £12.06/6 fittings = £2.01

Fittings with three ends:
    15 mm: 5 at £2.72   £13.60
    22 mm: 1 at £3.62   3.62

              £17.22/6 fittings = £2.87

Soil, waste and ventilation piping

Table 14.2 shows ranges of outputs for soil, waste and ventilation piping.

Table 14.2   *Soil, waste and ventilation piping fixed to masonry background: range of outputs per plumber in a squad in hours per metre and jointing sundries*

| | Cast iron pipework | | uPVC pipework | |
	50 mm	100 mm	50 mm	100 mm
Pipe	0.25–0.50	0.35–1.00	0.27–0.33	0.40–0.50
Bend	0.33–0.56	0.75–1.06	0.20–0.39	0.40–0.72
Single branch	0.50–0.76	0.95–1.50	0.33–0.45	0.60–1.00
Offset 150 mm		0.56–0.65		0.39–0.40
Offset 300 mm		0.80–1.06		0.45–0.72
Lead per joint	0.91–1.00 kg	1.50–1.81 kg	—	—
Yarn per joint	0.10 kg	0.10–0.20 kg	—	—

Example 14.5

> 100 mm nominal bore cast iron spigot and socket jointed soil and ventilation piping with staved lead joints and two piece holderbats at 1.8 m centres fixed to masonry.   m

Materials

Pipe in 1.8 m length at £10.02	£10.02
Holderbat, one per length at £2.50	2.50
Lead per joint: 1.5 kg at 83p/kg	1.25
Yarn per joint: 0.15 kg at 48p/kg	0.07
	13.84
Waste 5%	0.69

Labour

Method A: fix length of pipe including joint and holderbat:

1.08 hours at £7.65	8.26	
Method B:		
Fix holderbat, hang pipe: 1.8 m at 0.3 hours/m at £7.65	£4.13	
Run joint: 0.5 hours at £7.65	3.83	
	7.96	
Including labour by method A		8.62
		22.79
Cost per metre		12.66
Profit and oncost 20%		2.53
Rate per metre		£15.19

Example 14.6

Extra over 100 mm nominal bore cast iron soil pipe for 92.5° bend.   Nr

'Equivalent lengths' of cast iron pipe for bend calculations are normally available from manufacturers.

Materials	
Bend	£9.51
Joint: lead as pipe	1.25
yarn as pipe	0.07
	10.83
Waste 5%	0.54
	11.37
Labour	
Plumber 0.9 hours at £7.65	6.89
	18.26
Deduct cost (Example 14.5) of equivalent length of pipe incl. holderbat (joints cost taken with fitting): 0.254 × £12.66	3.22
	15.04
Profit and oncost 20%	3.01
Each	£18.05

Example 14.7

Extra over 100 mm nominal bore cast iron soil pipe for single branch 92.5°   Nr

Materials	
Branch	£14.76
Joints 2: lead as pipe, 2 at £1.25	2.50
yarn as pipe, 2 at £0.07	0.14
	17.40
Waste 5%	0.87
	18.27
Labour	
Plumber 1.50 hours at £7.65	11.48
	29.75
Deduct cost equivalent length pipe (see Example 14.6): 0.386 × £12.66	4.89
	24.86
Profit and oncost 20%	4.97
Each	£29.83

Example 14.8

| 102 mm bore uPVC soil and ventilation piping with O-ring sealed spigot and socket joints, standard fixings at 2 m centres plugged and screwed to masonry. | m |

Materials

Pipe, 4 m length	£13.14
Standard fixing: 2 at 62p	1.24
Screws: £4.50/100, 4 at 4.5p	0.18
Plugs: £2.30/100, 4 at 2.3p	0.09
	14.65
Waste 5%	0.73
	15.38

Labour

Plumber 4 m at 0.4 hours/m at £7.65	12.24
	27.62
Cost per metre	6.81
Profit and oncost 20%	1.38
Rate per metre	£8.29

Example 14.9

| Plastics solvent welded boss connection for 32/38 mm waste pipe to uPVC soil pipe. | Nr |

Materials

Connection	£1.28
Waste 2.5%	0.03
	1.31

Labour

Cut hole, solvent weld boss: 0.45 hours at £7.65	3.44
	4.75
Profit and oncost 20%	0.95
Each	£5.70

Example 14.10

| Plastics waste pipe with push fit O-ring sealed joints, fixed at given centres with plastics saddles, screwed to timber. | m |

Pipe diameter	32 mm	38 mm	50 mm
Materials			
Pipe in 4 m lengths	£4.26	£5.24	£7.68
Saddles: 4/0.75 say 5 at 32p	1.60	1.60	
4/1.00 say 4 at 32p			1.28
Screws: £2.00/100, 10 at 2p	0.20	0.20	
8 at 2p			0.16
Coupling in running length	0.42	0.48	0.60
	6.48	7.52	9.72
Waste 5%	0.32	0.38	0.49
	6.80	7.90	10.21
Labour (inc. joints in running length): 4 m at 0.3 hours/m at £7.65	9.18	9.18	9.18
	15.98	17.08	19.39
Cost per metre	4.00	4.27	4.85
Profit and oncost 20%	0.80	0.85	0.97
Rate per metre	£4.80	£5.12	£5.82

Example 14.11

> Extra over plastics waste pipe 65 mm nominal diameter with push fit O-ring sealed joints for fittings.

Pipe diameter	32 mm	38 mm	50 mm
Fittings with two ends (bent coupling)	£0.74	£0.83	£1.25
Waste 5%	0.04	0.04	0.06
	0.78	0.87	1.31
Labour (including joints): 0.2 hours at £7.65	1.53	1.53	1.53
	2.31	2.40	2.84
Profit and oncost 20%	0.46	0.48	0.57
Rate each	£2.77	£2.88	£3.41
Fittings with three ends (swept tee)	£1.10	£1.35	£2.35
Waste 5%	0.06	0.07	0.12
	1.16	1.42	2.47
Labour (including joints): 0.33 hours at £7.65	2.53	2.53	2.53
	3.69	3.95	5.00
Profit and oncost 20%	0.74	0.79	1.00
Each	£4.43	£4.74	£6.00

The first thing to note in this example is that the cost of the equivalent length of pipe has not been deducted. This length is relatively short and is not considered worth the effort. However, it is on the borderline and some might argue that a deduction should be made.

The second thing is that the rates for fittings would have to be proportioned out in the same way as those for the copper water supply piping in Example 14.4. Also, although the rates for plastics fittings are generally higher, the difference between rates is not as great as in copper capillary fittings, and so proportioning need not have such potentially serious financial consequences.

Example 14.12

> 22 mm uPVC overflow pipe with solvent welded joints, fixed at 600 mm centres with plastics clips screwed to timber.   m

| | 22 mm pipe | Extra over pipe for fittings with: | | |
		2 ends (bend)	3 ends (tee)	Tank conn.
Materials				
Pipe in 4 m lengths	£1.84	£0.44	£0.38	£0.60
Clips: 4/0.6 say 7 at 6p	0.42			
Joints in running length say every 6 m				
4/6 m = 0.66 at 30p	0.20			
	2.46	0.44	0.38	0.60
Waste 5%	0.12	0.02	0.02	0.03
Labour (including joints)				
4 m at 0.22 hours/m at £7.65	6.73			
0.15 hours at £7.65		1.15	1.15	
0.20 hours at £7.65				1.53
	9.31	1.61	1.55	2.16
Cost per metre	2.33			
Profit and oncost 20%	0.47	0.32	0.31	0.43
Rate per metre/fitting	£2.80	£1.93	£1.86	£2.59

Note that the cost of the equivalent length is once again ignored.

Rainwater goods

Table 14.3 shows ranges of outputs for rainwater goods. Waste is normally 5 per cent, but for excessive short length work allow at least 10 per cent.

Example 14.13

> 102 mm half round cast iron guttering, BS 640, with bolted joint sealed with mastic, hung on and including galvanized steel fascia brackets at 900 mm centres screwed to timber   m

Materials

Gutter in 1.8 m lengths	£10.62
Fascia brackets: 2 at 90p	1.80
Screws: £3.00/100, 4 at 3p	0.12
Mastic in joint: 0.25 kg at £2.30/kg	0.58
M6 gutter bolt and nut: 1 at 8p	0.08
	13.20
Waste 5%	0.66

Labour

Fixing brackets, hanging gutter, jointing: 1.8 m at 0.55 hours/m at £7.65	7.57
	21.43
Cost per metre	11.91
Profit and oncost 20%	2.38
Rate per metre	£14.29

Table 14.3 *Rainwater goods, piping fixed to masonry background, gutters to timber background: range of outputs per plumber in a squad in hours per metre and jointing sundries*

	100–120 mm gutter and fittings	65 mm pipe and fittings	100 mm pipe and fittings
*Cast iron goods**			
Straight length	0.30–0.55	0.18–0.54	0.41–0.46
Centre drop	0.25–0.45	—	—
Bend/angle	0.25–0.45	0.18–0.50	0.60–0.81
Stop end	0.12–0.16	—	—
Shoe	—	0.25–0.33	0.38–0.41
uPVC goods			
Straight length	0.26–0.44	0.18–0.54	0.41–0.46
Centre drop	0.25–0.45	—	—
Bend/angle	0.16–0.26	0.15–0.45	0.32–0.48
Stop end	0.10–0.16	—	—
Shoe	—	0.15–0.32	0.24–0.32
Mastic per joint†	0.15–0.20 kg	0.07 kg	0.10 kg

* For heavy gauge cast iron, i.e. 6 mm thick and over, times hours by 1.2 to 1.4.
† This only applies to cast iron goods. Plastics gutters have an integral seal. Down pipes are normally loose jointed; if they have to be sealed, an O or a D ring seal would be part of the pipe as supplied.

The labour constant of 0.55 hours/m for all work in connection with the gutter is not the only way to tackle this part of the rate. Some estimators prefer to keep constants for three distinct operations, namely fixing a bracket (times two in this instance), handling a length of gutter and making the joint. A very few allocate labour to a fourth category: taking delivery. The latter is a carryover from the days when every piece of material was manhandled off a truck; cast iron rainwater goods took a long time to unload. With palletized goods and mechanical off-loading on site, using either lorry mounted cranes or fork-lift trucks, this is no longer needed. The labour constant for hanging the gutter includes fetching it from storage on site, which is now the only piece of physical carrying done; even this has been reduced on larger sites.

Example 14.14

Extra over 102 mm cast iron gutter for double socketed fittings.	Nr

	Centre drop	Angle	Stop end
Materials			
Fitting	£4.18	£4.02	£1.36
Mastic in joint as gutter:			
2 × 0.25 kg at £2.30/kg	1.15	1.15	
1 × 0.25 kg at £2.30/kg			0.58
	5.33	5.17	1.94
Waste 5%	0.27	0.26	0.10
Labour			
Jointing and fixing: 0.40 hours at £7.65	3.06	3.06	
0.14 hours at £7.65			1.07
	8.66	8.49	3.11
Deduct cost of equivalent length (mm) of gutter at £11.91/m	(240)	(215)	(125)*
	2.86	2.56	1.49
	5.80	5.93	1.62
Profit and oncost 20%	1.16	1.19	0.32
Each	£6.96	£7.12	£1.94

* Some patterns of stop end do not displace any gutter and so no equivalent length is deductible.

Example 14.15

65 mm cast iron spigot and socket rainwater pipes, loose jointed, with two-piece holdfasts to masonry at 1.8 m centres.	m

Materials	
Pipe in 1.8 m length	£21.60
Holdfast	2.00
	23.60
Waste 5%	1.18
Labour	
Fixing holdfast, erecting length of pipe: 0.3 hours at £7.65	2.29
	27.07

Cost per metre	15.04
Profit and oncost 20%	3.01
Rate per metre	£18.05

Example 14.16

Extra over 65 mm cast iron loose jointed rainwater pipe for fittings.	Nr

	Bend	Offset 150 mm	Shoe
Materials			
Fitting	£6.06	£9.59	£10.23
Holdfast	2.00	2.00	2.00
	8.06	11.59	12.23
Waste 5%	0.40	0.58	0.61
Labour			
Holdfast and fixing: 0.45 hours at £7.65	3.44	3.44	
0.33 hours at £7.65			2.52
	11.90	15.61	15.36
Deduct cost of equivalent length (mm) of pipe at £15.04/m	(215)	(350)	(165)
	3.23	5.26	2.48
	8.67	10.35	12.88
Profit and oncost 20%	1.73	2.07	2.58
Each	£10.40	£12.42	£15.46

Example 14.17

112 mm uPVC half round gutter, BS 4567, on standard fascia brackets at 900 mm centres, screwed to timber.	m

Materials	
Gutter in 4 m lengths with integral joints	£4.04
Fascia brackets: 4/0.9, say 5 at 31p	1.55
Screws: £3.00/100, 5 × 2 at 3p	0.30
	5.89
Waste 5%	0.30
Labour	
Fixing brackets, hanging gutter, jointing: 4 m at 0.35 hours/m at £7.65	10.71
	16.90
Cost per metre	4.23
Profit and oncost 20%	0.85
Rate per metre	£5.08

Example 14.18

Extra over 112 mm uPVC gutter for the following fittings (fittings are usually double socketed where appropriate).		Nr	

	Centre drop	Angle	Stop end
Materials			
Fitting	£0.98	£1.14	£0.55
Fascia bracket	0.31	0.31	
Screws: 2 at 3p	0.06	0.06	
	1.35	1.51	0.55
Waste 5%	0.07	0.08	0.03
Labour			
Fixing brackets, jointing: 0.4 hours at £7.65	3.06	3.06	
0.1 hours at £7.65			0.77
	4.48	4.65	1.35
Deduct cost of equivalent length (mm) of gutter at £4.23/m	(200)	(200)	(nil)
	0.85	0.85	nil
	3.63	3.80	1.35
Profit and oncost 20%	0.73	0.76	0.27
Each	£4.36	£4.56	£1.62

Equivalent lengths for PVC rainwater goods are obtained from manufacturers' catalogues or by direct measurement of samples; there are no standard sizes. Stop ends are generally made in two varieties: those ftting over the end of the gutter, and those fitting inside the socket joint of the length of gutter. In either instance there is no saving in the overall length of gutter used.

Example 14.19

68 mm uPVC rainwater pipe BS 4576, loose jointed with PVC pipe clips at 900 mm centres, plugged and screwed to masonry.	m

Materials	
Length of pipe 4 m	£3.68
Pipe clips: 4/0.9, say 5 at 41p	2.05
Screws: £3.00/100, 5 × 2 at 3p	0.30
Plugs: £2.30/100, 5 × 2 at 2.3p	0.23
	6.26
Waste 5%	0.31
Labour	
Fixing clips and erecting pipe: 4 m at 0.3 hours/m at £7.65	9.18
	15.75

Cost per metre	3.94
Profit and oncost 20%	0.79
Rate per metre	£4.73

Example 14.20

Extra over uPVC loose jointed rainwater pipe >65 mm diameter for the following fittings.	Nr

	Bend	Offset 150 mm	Shoe
Materials			
Fitting	£0.65	£0.73	£0.73
Clip	0.41	0.41	0.41
Screws: 2 at 3p	0.06	0.06	0.06
Plugs: 2 at 2.3p	0.05	0.05	0.05
	1.17	1.25	1.25
Waste 5%	0.06	0.06	0.06
Labour			
0.4 hours at £7.65	3.06	3.06	
0.3 hours at £7.65			2.29
	4.29	4.37	3.60
Deduct cost of equivalent length (mm) of pipe at £3.94/m	(215)	(350)	(165)
	0.85	1.38	0.65
	3.44	2.99	2.95
Profit and oncost 20%	0.69	0.60	0.59
Each	£4.13	£3.59	£3.54

Plastic water supply piping

Table 14.4 shows ranges of outputs for plastics water pipe.

Table 14.4   *Plastics water pipe: range of outputs per plumber in a squad in hours/m*

	Nominal bore			
	$\frac{1}{2}$ in (12.5 mm)	$\frac{3}{4}$ in (19 mm)	1 in (25 mm)	$1\frac{1}{4}$ in (32 mm)
Black polythene type 32 water pipe to BS 1972 for above ground use				
Clipped to timber	0.10–0.33	0.12–0.48	0.14–0.60	0.17–0.70
Clipped to masonry	0.20	0.24	0.28	0.34
Blue polythene water pipe to BS 6572 for underground use				
Laid in trench	0.01–0.06	0.013–0.09	0.015–0.12	0.02–0.15
Any fitting	0.50	0.60	0.70	0.80

Example 14.21

½ in nominal bore Class B polythene piping, to BS 1972, fixed with plastics clips at 600 mm centres screwed to timber background, including joints in the running length.	m

Materials

Pipe in 150 m coils	£42.00
Clips*: 150/0.6, say 250 at 3p	7.50
Screws: £2.00/100, say 250 at 2p	5.00
Joints: take 1 every 30 m on average:	
Type A couplings: say 5 at £1.08	5.40
Liners for couplings: 10 at £0.14	1.40
	61.30
Waste 5%	3.07

Labour

Plumber 150 m at 0.1 hours/m at £7.65	114.75
	179.12
Cost per metre	1.19
Profit and oncost 20%	0.24
Rate per metre	£1.43

*Clips are one-piece snap type with single screw fixing.

Example 14.22

Extra over polythene pipe ⩽ 65 mm diameter for type A fittings with ends as follows.	Nr

	Two ends (reducer)	Three ends (tee)
Materials		
Fitting	£0.94	£3.78
Pipe liners at 14p – 1 per end	0.28	0.42
	1.22	4.20
Waste 5%	0.06	0.21
Labour		
0.6 hours at £7.65	4.59	
0.9 hours at £7.65		6.89
	5.87	11.30
Profit and oncost 20%	1.17	2.26
Each	£7.04	£13.56

Fittings on polythene pipework are expensive. The estimator faced with the task of proportioning the rates for fittings with two and three ends must exercise considerable care, lest he underprice or overprice the job.

Example 14.23

Full value rate for type A compression fittings to polythene pipework.	Nr	

	Connection + backnut	SD valve
Materials		
Fitting	£1.88	£7.52
Pipe liners at 14p	0.14	0.28
	2.02	7.80
Waste 5%	0.10	0.39
Labour		
Plumber 0.6 hours at £7.65	4.59	4.59
	6.71	12.78
Profit and oncost 20%	1.34	2.56
Each	£8.05	£15.34

Sanitary fittings

Table 14.5 shows outputs for sanitary fittings. The wide range of outputs in the table seems to depend on whether or not the estimator includes traps, taps, wastes, plugs, chains and chain stays in the overall hours (see the examples which follow).

Table 14.5 *Sanitary fittings: range of outputs per plumber in a squad in hours per unit*

Fitting	Hours
Close coupled WC	1.70–2.35
Low level WC	2.25–4.00
Pedestal wash hand basin	2.05–3.10
Wash hand basin on brackets	1.75–2.50
Rectangular bath: cast iron	3.00–7.00
acrylic	2.65–6.00
Bidet with plain rim	1.95–2.25
Shower slab, control and shower head	6.00
Urinals: stall (per metre)	3.00–5.00
bowl	0.90–2.00
Vitreous china sink	1.25–2.75
Stainless steel sink	1.90–4.50

Example 14.24

White vitreous china washdown WC, BS 1213, with low level cistern, flushing pipe, plastic seat and cover and push fit connection to uPVC soil pipe (timber floor, masonry wall).	Nr

Materials

WC pan	£26.30
Cistern	21.75
Flush pipe	0.90
Seat and cover	19.46
Screws: £3.00/100, 6 at 3p	0.18
Plugs: £2.30/100, 4 at 2.3p	0.09
Connector	2.60
	71.28
Waste 2.5%	1.78

Labour

Plumber 3.5 hours at £7.65	26.78
	99.84
Profit and oncost 20%	19.97
Each	£119.81

The connection for the incoming cold water supply to the cistern should be given in the bill as a separate full value item. So too should the tap connectors for the sanitary fittings in the examples which follows.

Example 14.25

White vitreous enamel pressed steel bath, BS 1390, complete with CP bath mixer taps, CP waste, plastic trap with combined overflow, plug, CP chain and chain stay integral with overflow fitting.	Nr

Materials

Bath	80.78
Mixer set	32.79
Waste 38 mm with combined overflow and chain stay	3.47
Bath trap	1.84
Sealing washers at waste	0.20
	119.08
Waste 2.5%	2.98

Labour

Plumber 3 hours at £7.65	22.95
	145.01
Profit and oncost 20%	29.00
Each	£174.01

Example 14.26

> White vitreous china wash hand basin, BS 1188 and BS 3402, complete with white vitreous china pedestal, plastics bottle trap, 32 mm CP waste, plug, chain stay and two $\frac{1}{2}$ in CP taps.　　　Nr

Materials
Basin	£18.54
Pedestal	18.19
Waste fitting	2.64
Bottle trap	1.34
Plug, chain and chain stay	1.25
Sealing washers at waste	0.20
Screws: £3.00/100, 4 at 3p	0.12
CP taps: 2 at £6.00	12.00
Plastics washers at taps	0.12
	54.40
Waste 2.5%	1.36

Labour
Plumber 3 hours at £7.65	22.95
	78.71
Profit and oncost 20%	15.75
Each	£94.46

Water tanks and cylinders

Table 14.6 shows ranges of outputs for water and gas equipment. Control heating systems are dealt with in the next section.

The reader may well find outputs per unit in the region of 5–7 hours per tank or hot water cylinder. Outputs of this magnitude include cutting all holes and fitting overflows and supply pipes in and out of the tank, as well as ball valves, standing wastes etc. The number of these is obviously an average.

Example 14.27

> A galvanized open top cistern with loose lid, BS 417 Part 2, 213 litres actual capacity, with 32 mm standing waste and 22 mm ballcock with tail for copper and 22 mm silencer pipe.　　　Nr

Materials
Tank and lid	£46.61
Ballcock	9.00
Brass waste and tapered plug	1.20
Standing waste 32 mm copper pipe, 600 mm long	4.10
Silencer pipe 22 mm copper, 750 mm long, one bend, braze to ballcock	1.32
	62.23
Waste 2.5%	1.56

Labour

Setting tank: 1 hour at £7.50	7.50
Cut holes for supply piping: say 4 × 0.12 hours at £7.65	3.67
Soft solder 32 mm pipe to tapered plug, flare end and fit in position: 0.3 hours at £7.65	2.29
Long splay cut on 22 mm pipe, one form bend, soft solder to ballcock body and fit ballcock: 0.50 hours at £7.65	3.83
	81.08
Profit and oncost 20%	16.22
Each	£97.30

Generally the connections for joining pipes to tanks, cylinders, boilers etc. are measured as full value separate items in a bill of quantities, so that only necessary hole cutting is taken with the appliance.

Table 14.6   *Tanks, cylinders and boilers: range of outputs per plumber in a squad in hours per unit*

Fitting	Hours
Water tanks and cylinders	
PVC tank, BS 4213: 4 gal (18 litres)	0.50–0.60
50 gal (227 litres)	1.00–1.25
Direct, indirect and primatic HW cylinders, 30 gal (136 litres)	0.50–0.70
Combination tank/HW cylinder	0.90–1.00
Cutting holes in tanks and cylinders	0.12
Gas boilers for small domestic CH, balanced flue	
Floor mounted	3.75–4.10
Wall mounted	3.75–4.10
Multipoint boiler	2.50
Electric immersion: top entry	0.10–0.15
side entry inc. cutting hole	1.00–1.25
Radiators up to 2.00 m long and 1.00 m high	
Per hanging bracket*	0.075–0.15
Fitting air bleed valve, blank plug, lock shield valve and control valve	0.50–0.75

* Radiators up to 1 m wide are generally hung on two brackets and up to 2 m wide on three brackets.

Example 14.28

Indirect copper hot water storage cylinder, BS 1566 Part 1, 136 litres capacity, complete with threaded connection for top entry immersion heater.	Nr

Materials

Cylinder	£50.57
Waste 2.5%	1.26

Labour

Setting 0.6 hours at £7.65	4.59
Fitting immersion 0.1 hours at £7.65	0.77
	57.19
Profit and oncost 20%	11.44
Each	£68.63

Holes for pipe connections are never cut in an HW cylinder; the manufacturer supplies the cylinder with suitable male or female threaded bosses to which connectors are fastened for the supply pipe.

Immersion heater connections are now generally made through a special boss in the top of the cylinder. However, it is still possible to have an immersion connected through an oval hole cut in the side of the cylinder; in that case the time taken to fit the immersion heater would be increased to approximately 1 hour. It is assumed that the immersion heater will be given as a separate item in the bill. On occasions the item for the supply of the immersion heater is given in the electrical bill, and only fitting and attendance are given in the plumbing work.

Central heating systems

Selected outputs are given in Table 14.6.

Example 14.29

Floor mounted gas fired boiler, approximately 8–12 kW output, with a balanced flue.	Nr

Materials

Boiler and flue	£285.00
Sealing flue to outer wall: say	2.50
	287.50
Waste 2.5%	7.19

Labour

Fitting flue through wall, cutting hole and setting boiler: 4 hours at £7.65	30.60
	325.29
Profit and oncost 20%	65.06
Each	£390.35

Connections for pipework, pump, controller and external thermostats are usually measured – and priced – separately. The boiler in this example includes a factory fitted and set safety valve and limiting thermostats. Boiler packages are now available where pump and controller are not only supplied but plumbed into the casing. Even a pressure vessel can be supplied plumbed in. In such cases only a cold feed, an open vent and the flow and return pipes are required, plus electrical and gas connections.

Example 14.30

Microbore copper piping, BS 2871 Table Y, including capillary joints in the running length and clips at 500 mm maximum centres, nailed to timber or masonry		m	

Pipe diameter	8 mm	10 mm	12 mm
Materials			
Pipe (in 25 m coils), per 100 m	£78.40	£98.10	£122.40
Joints in running length:			
say every 6 m: say 17 at 26p	4.42	4.42	
17 at 29p			4.93
Clips: £3.00/100, say 200	6.00	6.00	6.00
	88.82	108.52	133.33
Waste 5%	4.44	5.43	6.67
Labour			
100 m at 0.1 hours/m at £7.65	76.50	76.50	76.50
	169.76	190.45	216.50
Cost per metre	1.70	1.91	2.17
Profit and oncost 20%	0.34	0.38	0.43
Rate per metre	£2.04	£2.29	£2.60

The inclusion of joints in the running length at 6 m intervals may seem peculiar, as it has been stated that the coils are 25 m long. Why not a joint every 25 m? The reason is that it is unlikely that the plumber will be able to lay more than 6 m at a time on *average* because of obstructions. The pipe is fully annealed, and theoretically it should be possible to draw the whole 25 m through a series of obstructions. In practice the pipe hardens and can kink and fracture, so drawing in is kept to a minimum. It would also be theoretically possible to apply heat to the pipe and anneal any portions which become work hardened. However, heating costs fuel and time, and this cost is far greater than that of the odd additional joint (three in this case). Heating also carries a fire hazard, and the inner surface of the pipe can become oxidized if the plumber keeps annealing it.

Example 14.31

22 mm brass one-way flow manifold with capillary joint to copper and 4 × 10 mm side entries, type A compression joints to copper.	Nr

Materials	
Manifold	£3.29
Waste 2.5%	0.08

Labour

Capillary joint each end 2×0.28	0.56	
Compression joints 4×0.15	0.60	
	1.16 hours at £7.65	8.87
		12.24
Profit and oncost 20%		2.45
Each		£14.69

Example 14.32

22 mm brass one-way flow manifold with 4×8 mm linear flow connections, all capillary joints.	Nr

Materials	
Manifold	£2.72
Waste 2.5%	0.07
Labour	
Plumber 0.28 hours at £7.65	2.14
	4.93
Profit and oncost 20%	1.00
Each	£5.93

Example 14.33

Pressed steel pre-finished radiators hung on brackets supplied, screwed and plugged to masonry; two single entry control valves with tails for copper pipe.	Nr

	Single panel		Double panel	
	800 w $\times 440$ h	1280 w $\times 740$ h	800 w $\times 440$ h	1280 w $\times 740$ h
	0.50 kW	1.26 kW	0.78 kW	1.90 kW
Materials				
Radiator	£12.55	£33.90	£24.65	£58.70
2 CP valves at £2.82	5.64	5.64	5.64	5.64
Screws: £3.00/100: 6 at 3p	0.18		0.18	
9 at 3p		0.27		0.27
Plugs: £2.30/100: 6 at 2.3p	0.14		0.14	
9 at 2.3p		0.21		0.21
	18.51	40.02	30.61	64.82
Waste 2.5%	0.46	1.00	0.77	1.62

Labour
Fixing brackets:

2 × 0.1 hours at £7.65	1.53		1.53	
3 × 0.1 hours at £7.65		2.29		2.29

Fitting valves, blanking plug and air cock,
hang radiator inc. 2 conns to copper:

0.6 hours at £7.65	4.59	4.59	4.59	4.59
	25.09	47.90	37.50	73.32
Profit and oncost 20%	5.02	9.58	7.50	14.66
Each	£30.11	£57.48	£45.00	£87.98

Example 14.34

50 mm thick glass fibre, polythene clad, insulation set for rectangular water storage cistern size 600 × 450 × 475 mm, fixed with bands and clips supplied.   Nr

Materials

Insulation set	£2.30
Waste 5%	0.12

Labour

Plumber 0.75 hours at £7.65	5.74
	8.16
Profit and oncost 20%	1.63
Each	£9.79

Example 14.35

80 mm thick glass fibre, PVC coated, cambric clad, segmented insulation set for hot water storage cylinder size 400 dia-meter × 1050 mm high, fixed with bands and clips supplied.   Nr

Materials

Insulation set	£3.45
Waste 5%	0.17

Labour

Plumber 0.33 hours at £7.65	2.52
	6.14
Profit and oncost 20%	1.23
Each	£7.37

Example 14.36

Pre-formed foamed plastic pipe insulation, 19 mm wall thickness, with continuous self-fastening edge.		m	

Pipe diameter	15 mm	22 mm	28 mm
Materials			
Insulation per metre	£0.69	£0.84	£0.92
Waste 5%	0.03	0.04	0.05
Labour			
Plumber 0.08 hours at £7.65	0.61	0.61	0.61
	1.33	1.49	1.58
Profit and oncost 20%	0.27	0.30	0.32
Rate per metre	£1.60	£1.79	£1.90

15

Linings, partitions and surface finishes

SMM7 Sections K10, K31, K41, M10, M20, M30, M40

Plasterboard

Plasterboard is available in various types for different uses, and in various widths, lengths and thicknesses. The main manufacturer of plasterboard in Britain is British Gypsum Limited. Therefore that company's products are described, as they are those most likely to be encountered on a day-to-day basis.

Gyproc wallboard and plank

Gyproc wallboard is a dry lining plasterboard consisting of an aerated gypsum core encased in a durable paper liner, suitable for application to internal surfaces. The boards have one face of an ivory coloured finish for use where decoration will be applied direct, that is where joints are filled and taped, and the other face of a grey finish which can be coated with plaster.
   There are three types of edge profile for differing joint requirements:

● Tapered edge for smooth seamless joints
● Square edge for cover strip jointing or plastering
● Bevelled edge for V jointing.

Wallboards are manufactured in thicknesses of 9.5 mm and 12.7 mm and in widths of 600, 900 and 1200 mm, with the exception of square edge boards which are produced in widths of 900 and 1200 mm only. The most commonly used lengths of board are 1800 and 2400 mm, although boards are produced up to 3500 mm long.
   Gyproc plank is a dry lining board manufactured to similar specifications as the wallboard. The main differences are that plank is produced either with both faces in grey or with one grey and one ivory, and that it is only available in 600 mm widths and in one thickness of 19.00 mm.

Gyproc Duplex plasterboards

Gyproc Duplex plasterboards combine the same qualities as ordinary plasterboards with a vapour resistant lining on one face for compliance with current Building Regulations.

Thistle baseboard and Gyproc lath

Thistle baseboard is a lining board for use as a base for gypsum plaster to timber ceiling, partition and wall members. It has an aerated core with grey paper surfaces, and is square on all edges. It is normally supplied in 1200 × 900 mm boards.
   Gyproc lath is used in the construction of a suspended ceiling system which incorporates rigid support brackets, enabling the application of plaster bonding and finishing coats.

Table 15.1 *Range of labour outputs for plasterboards*

Description	Application	Thickness (mm)	Squad hours per 100 m²
Wallboard	Walls	9.5	10–15
		12.7	12–18
	Ceilings	9.5	11–17
		12.7	13–20
Plank	Walls	19.0	15–20
	Ceilings	19.0	17–23
Duplex wallboard	Walls	9.5	10–15
		12.7	12–18
Baseboard	Walls	9.5	10–15
	Ceilings	9.5	11–17
Gyproc lath	Walls	9.5	11–16
		12.7	13–19
	Ceilings	9.5	12–18
		12.7	15–21
Thermalboard	Walls	25.0	14–19
		50.0	17–23
	Ceilings	25.0	16–22
		50.0	19–26
Fireline board	Walls	12.7	13–19
	Ceilings	12.7	15–21
Filling joints and scrim	Walls		6–10
	Ceilings		8–12
Filling, taping joints, slurry coat	Walls		8–12
	Ceilings		10–14

Gyproc thermal board

Gyproc thermal board is a laminated board composed of a Gyproc wallboard bonded to a layer of expanded polystyrene. This gives a lining board with good thermal insulation. It is manufactured in 25, 32, 40 and 50 mm thicknesses. It has a grey paper surface with square edges for plastering, or an ivory surface with tapered edges for filling and taping ready for decoration. The boards are also available with a vapour resisting membrane at the interface of the polystyrene and the wallboard.

Gyproc Fireline board

Gyproc Fireline board is a gypsum plasterboard with the addition of glass fibre and vermiculite in the core, resulting in improved fire protection properties. The boards can be used as a lining to walls, partitions and ceilings and as a casing to steel beams and columns, increasing the structure's fire resistance.

Boards have one face of ivory coloured finish with tapered or square edges, and one grey face. Fireline boards are manufactured in one thickness of 12.7 mm and in widths of 600, 900 and 1200 mm. The 600 mm board is produced in square edge only. The most commonly used lengths of board are 1800 and 2400 mm, although boards are produced up to 3600 mm long.

Table 15.1 gives a range of labour outputs to fix plasterboards. The hours are based on a squad consisting of 2 tradesmen and 1 labourer.

Example 15.1

> Linings to walls on timber base: Gyproc wallboard 9.5 mm thick, tapered edge; butt jointed, filled with joint filler, tape and joint finish; spot filling; coat whole surface with slurry coat of joint finish; fixing with galvanized nails; to walls 2.1–2.4 m high.   m

Materials

Cost of Gyproc wallboard at £153 per 100 m^2	£153.00
Take delivery and stack 100 m^2: 0.75 hours at £5.30	3.97

Galvanized plasterboard nails to fix 1200 × 2400 mm board at 150 mm centres:
  To timber members at 600 mm centres: $3 \times [(2400/150)+1] = 51$
  To dwangs/noggings at 1200 mm centres: $3 \times [(1200/150)+1-3] = 18$
  Nr nails per sheet: $51 + 18 = 69$
  Nr nails per 100 m^2: $69 \times 100/(1.2 \times 2.4) = 2395.83$

Cost per 100 m^2: 2396 at 110/kg at £2.50 kg	54.45

Joint tape to half perimeter of board: $1.2 + 2.4 = 3.6$ m
Length of tape per 100 m^2: $3.6 \times 100/(1.2 \times 2.4) = 125$ m

Cost of tape per 100 m^2: 125 m at £0.02/m	2.50
	213.92
Wastage 5%	10.70
	224.62
Joint filler inc. wastage: 22 kg per 100 m^2 at £0.30/kg	6.60
Joint finish inc. wastage: 33 kg per 100 m^2 at £0.44/kg	14.52
Slurry coat inc. wastage: 25 kg per 100 m^2 at £0.44/kg	11.00
Take delivery and stack at 40 nr 25 kg bags per hour:	
100 m^2 requires 80 kg: 0.08 hours at £5.30	0.42

Labour
Squad of 2 tradesmen and 1 labourer will fix 100 m^2 of wallboard in 12 hours:
  Labour cost per hour $(2 \times £6.10)+(1 \times £5.30) = £17.50$

Cost per 100 m^2: £17.50 × 12	210.00
Fill joints, tape, finish and slurry coat 100 m^2 will take squad 10 hours:	
Cost per 100 m^2: £17.50 × 10	175.00
	642.16
Profit and oncost 20%	128.43
Cost per 100 m^2	770.59

Cost per m² £7.71

However, in accordance with SMM7 the linings to walls are measured in linear metres; the height is given in stages of 300 mm. The estimator should determine from drawings the specific heights of the walls or partitions that are to be lined. The difference between say 2100 mm and 2400 mm will have a great effect on the competitiveness of the rate and ultimately the profitability, as shown by the following:

Walls 2.1 m high:
    rate per metre = cost per m² × wall height = £7.71 × 2.1 = £16.19
Walls 2.4 m high:
    rate per metre = cost per m² × wall height = £7.71 × 2.4 = £18.50

Example 15.2

> Linings to ceilings on timber base: Gyproc wallboard 9.5 mm thick, tapered edge; butt jointed, filled with joint filler, tape and joint finish; spot filling; coat whole surface with slurry coat of joint finish; fixing with galvanized nails; to ceilings not exceeding 3.5 m above floor level.   m²

Materials
Cost of Gyproc wallboard at £154 per 100 m²   £154.00
Take delivery and stack 100 m²: 0.75 hours at £5.30   3.97

Galvanized plasterboard nails to fix 1800 × 900 mm board at 150 mm centres:
    To timber members at 600 mm centres: 4 × [(900/150) + 1] = 28
    To dwangs/noggings at 900 mm centres: 2 × [(1800/150) + 1 − 4] = 18
    Nr nails per sheet: 28 + 18 = 46
    Nr nails per 100 m²: 46 × 100/(1.8 × 0.9) = 2839.51
Cost per m²: 2840 at 110/kg at £2.50 kg   64.55

Joint tape to half perimeter of board: 1.8 + 0.9 = 2.7 m
Length of tape per 100 m²: 2.7 × 100/(1.8 × 0.9) = 166.67 m
Cost of tape per 100 m²: 166.67 m at £0.02/m   3.33
                                                                               ——————
                                                                               225.85
Wastage 5%   11.29
                                                                               ——————
                                                                               237.14
Joint filler inc. wastage: 26 kg per 100 m² at £0.30/kg   7.80
Joint finish inc. wastage: 38 kg per 100 m² at £0.44/kg   16.72
Slurry coat inc. wastage: 25 kg per 100 m² at £0.44/kg   11.00
Take delivery and stack at 40 nr 25 kg bags per hour:
    100 m² requires 89 kg: 0.089 hours at £5.30   0.47

Labour
Squad of 2 tradesmen and 1 labourer will fix 100 m² of wallboard to ceilings in
  14 hours:
   Cost per 100 m²: £17.50 × 14   245.00
Fill joints, tape, finish and slurry coat 100 m² will take squad 12 hours:
   Cost per 100 m²: £17.50 × 12   210.00

                                               728.13
Profit and oncost 20%   145.63

Cost per 100 m²   873.76

Rate per m²   £8.74

Example 15.3

> Linings to walls on timber base: thistle baseboard 9.5 mm thick, butt
> jointed, joints filled with thistle board finish, reinforced with jute scrim;
> fixing with galvanized nails; to walls 2.1–2.4 m high.   m

Materials
Cost of thistle baseboard at £125 per 100 m²   £125.00
Take delivery and stack 100 m²: 0.75 hours at £5.30   3.97

Galvanized plasterboard nails to fix 1200 × 900 mm board at 150 mm centres:
   To timber members at 600 mm centres: $3 \times [(900/150) + 1] = 21$
   To dwangs/noggings at 900 mm centres: $2 \times [(1200/150) + 1 - 3] = 12$
   Nr nails per sheet: $21 + 12 = 33$
   Nr nails per 100 m²: $33 \times 100/(1.2 \times 0.9) = 3055.56$
Cost per 100 m²: 3056 at 110/kg at £2.50 kg   69.45

Joint scrim to half perimeter of board: $1.2 + 0.9 = 2.1$ m
Length of tape per 100 m²: $2.1 \times 100/(1.2 \times 0.9) = 194.44$ m
Cost of scrim per 100 m²: 194.44 m at £0.09/m   17.50

                                           215.92
Wastage 2.5%   5.40

                                         221.32
Carlite bonding coat filling joints inc. wastage: 30 kg per 100 m² at £0.12/kg   3.60
Take delivery and stack at 40 nr 25 kg bags per hour:
   100 m² requires 30 kg: 0.03 hours at £5.30   0.16

Labour
Squad of 2 tradesmen and 1 labourer will fix 100 m² of wallboard in 12 hours:
   Cost per 100 m²: £17.50 × 12   210.00
Fill joints, tape, and scrim 100 m² will take squad 6 hours:
   Cost per 100 m²: £17.50 × 6   105.00

                                         540.08
Profit and oncost 20%   108.02

Cost per 100 m²   648.10

Cost per m²   £6.48
Rate per m 2.4 m high (£6.48 × 2.4)   £15.55

Table 15.2   *Range of labour outputs for cement/sand floor screeds*

Description	Squad output
Cement and sand screed wood float:	
25 mm thick	0.16–0.20 hours/m²
50 mm thick	0.18–0.22 hours/m²
75 mm thick	0.22–0.26 hours/m²
Laying to slopes over 15° form horizontal per 25 mm thickness	0.05–0.10 hours/m²
Skirting 150 mm high with radius cove at floor	0.23–0.27 hours/m

Cement/sand screeds

Table 15.2 shows a range of labour outputs for laying cement/sand floor screeds. The hours are again based on a squad consisting of 2 tradesmen and 1 labourer.

Example 15.4

> Mortar, cement and sand 1:3; 50 mm work to floors; level and to falls only <15 degrees from horizontal; screeded.   m²

Materials

Washed sand per tonne	£7.50
Cement bagged delivered to site per tonne	61.50
Unload and stack: 1 hour/tonne at £5.30	5.30
	66.80
Materials cost per m³ = tonnes/m³ at cost/tonne:	
Cement: 1.28 at £66.80	85.50
Sand: 1.28 at £7.50	9.60
Mix by volume:	
1 m³ cement costs	85.50
3 m³ sand costs	28.80
	114.30
Allow 20% shrinkage: therefore add 25%	28.58
Cost 4 m³ of 1:3 mix	142.88
Cost 1 m³ of 1:3 mix	35.72
Cost per m² at 50 mm thick: £35.72 × 0.05	1.79
Wastage 5%	0.09

Labour
Squad of 2 tradesmen and 1 labourer will mix and lay 1 m² in 0.20 hours:
Cost per m²: £17.50 × 0.20   3.50
                                                                                     ―――
                                                                                     5.38
Profit and oncost 20%   1.08
                                                                                     ―――
Rate per m²   £6.46
                                                                                     ―――

Granolithic screeds

Table 15.3 shows a range of labour outputs for laying granolithic floor screeds. The hours are again based on a squad consisting of 2 tradesmen and 1 labourer.

Table 15.3   *Range of labour outputs for granolithic floor screeds*

Description	Squad output
Cement and granite screed steel trowel:	
25 mm thick	0.25–0.29 hours/m²
50 mm thick	0.30–0.34 hours/m²
75 mm thick	0.35–0.40 hours/m²
Laying to slopes over 15° from horizontal per 25 mm thickness	0.05–0.10 hours/m²
Skirting 150 mm high with radius cove at floor	0.28–0.33 hours/m

Example 15.5

> Granolithic, cement and granite 2:5; 25 mm work to floors; level and to falls only <15 degrees from horizontal; steel trowelled.   m²

Materials
Granite chippings per tonne delivered to site in bulk   £18.33

Cement bagged delivered to site per tonne   61.50
Unload and stack: 1 hour/tonne at £5.30   5.30
                                                                                     ―――
                                                                                     66.80
                                                                                     ―――

Materials cost per m³ = tonnes/m³ at cost/tonne:
Cement: 1.28 at £66.80   85.50
Granite chippings: 1.6 at £18.33   29.33

Mix by volume:
2 m³ cement at £85.50/m³   171.00
5 m³ chips at £29.33/m³   146.65
                                                                                    ――――
                                                                                    317.65
Allow 20% shrinkage: therefore add 25%   79.41
                                                                                    ――――
Cost 7 m³ of 2:5 mix   397.06
                                                                                    ――――

Cost 1 m³ of 2:5 mix	56.72

1 m³ 2:5 mix costs £56.72

Cost per m² at 25 mm thick: £56.72 × 0.025	1.42
Wastage 5%	0.07

Labour
Squad of 2 tradesmen and 1 labourer will mix and lay 1 m² in 0.27 hours:

Cost per m²: £17.50 × 0.27	4.73
	6.22
Profit and oncost 20%	1.24
Rate per m²	£7.46

Example 15.6

> Granolithic, cement and granite 2:5; 25 mm wide, 150 mm high skirting
> with 20 mm radius cove next to floor; on brickwork or blockwork base.   m

Materials
Cost 1 m³ of 2:5 mix (Example 15.5): £56.72

Cost 1 m skirting 150 mm high × 25 mm wide: £56.72 × 0.025 × 0.15	£0.21
Cost to form cove: say	0.01
Temporary grounds: say	0.01
	0.23
Wastage 5%	0.01

Labour
Squad of 2 tradesmen and 1 labourer will produce 1 m in 0.29 hours, including fixing
and removing temporary grounds:

Cost per m²: £17.50 × 0.29	5.08
	5.32
Profit and oncost 20%	1.06
Rate per metre	£6.38

Plastered and rendered coatings

Table 15.4 shows a range of labour outputs for applying plaster to walls. The hours are again based on a squad consisting of 2 tradesmen and 1 labourer.

Example 15.7

> Plaster to walls: Carlite; pre-mixed; floating coat of bonding 8 mm
> thick; finishing coat of finish 2 mm thick; 10 mm work to 9.5 mm
> plasterboard base to walls; steel trowelled; > 300 mm wide.   m²

Materials
Coverage of Carlite bonding 8 mm thick is 150 m² per 1000 kg
  Cost of Carlite bonding per 100 m²: 100/150 at £115 per tonne   £76.67
Coverage of Carlite finish 2 mm thick is 450 m² per 1000 kg
  Cost of Carlite finish per 100 m²: 100/450 at £95 per tonne   21.11
Take delivery and stack at 40 nr 25 kg bags per hour:
  100 m² requires 36 bags: 0.9 hours at £5.30   4.77
                                      102.55
Wastage 2.5%   2.56
                                      105.11

Labour
Squad of 2 tradesmen and 1 labourer will mix and apply 100 m² in 18 hours:
  Cost per 100 m²: £17.50 × 18   315.00
                                    420.11
Profit and oncost 20%   84.02
Cost per 100 m²   504.13

Rate per m²   £5.04

Table 15.4   *Range of labour outputs for plaster work*

Description	Squad hours per 100 m²
Float and finish coat on:	
Plasterboard	16–20
Brick or block	18–22
Concrete	16–20
Expanded metal lath	20–24
Two-coat render to brick or blockwork	18–22
Add to above labour outputs for work to:	
Ceilings	15%
Heights above 3.5 m per 1.5 m stage	5%
Widths not exceeding 300 mm	25%
Rounded angles: tradesman m/hour	15–20

Example 15.8

> Plaster to walls: Carlite; pre-mixed; floating coat of browning 11 mm thick; finishing coat of finish 2 mm thick; 13 mm work to walls on brickwork base; steel trowelled; >300 mm wide.   m²

Materials
Coverage of Carlite browning 11 mm thick is 140 m² per 1000 kg
   Cost of Carlite browning per 100 m²: 100/140 at £120 per tonne   £85.71
Coverage of Carlite finish 2 mm thick is 450 m² per 1000 kg
   Cost of Carlite finish per 100 m²: 100/450 at £95 per tonne   21.11
Take delivery and stack at 40 nr 25 kg bags per hour:
   100 m² requires 37 bags: 0.925 hours at £5.30   4.90

    111.72
Wastage 2.5%   2.79

    114.51

Labour
Squad of 2 tradesmen and 1 labourer will mix and apply 100 m² in 20 hours:
   Cost per 100 m²   350.00

    464.51
Profit and oncost 20%   92.90

Cost per 100 m²: £17.50 × 20   557.41

Rate per m²   £5.57

Example 15.9

> Plaster to isolated columns: Carlite; pre-mixed; floating coat of
> bonding 8 mm thick; finishing coat of finish 2 mm thick; 10 mm work
> to isolated columns on concrete base; steel trowelled; < 300 mm wide.   m

Materials
Cost 100 m² materials inc. wastage (Example 15.7)   £105.11

Labour
Squad of 2 tradesmen and 1 labourer will mix and apply 100 m² in 20 hours:
   Cost per 100 m²: £17.50 × 20   350.00
Additional labour for width < 300 mm: say 25%   87.50

Cost per 100 m²   542.61

Cost per m²   5.43

Assume 150 mm wide columns
Cost per metre for width 150 mm: £5.43 × (150/1000)   0.81
Profit and oncost 20%   0.16

Rate per metre   £0.97

Example 15.10

> Render: cement and sand 1:3; first coat 10 mm thick; finishing coat
> 3 mm thick; steel trowelled; 13 mm work to walls on brickwork or
> blockwork base; < 300 mm wide.   m

Materials
Cost 1 m³ of 1:3 mix (Example 15.4): £35.72
Cost per m² at 13 mm thick: £35.72 × 0.013   £0.46
Wastage 5%   0.02

Labour
Squad of 2 tradesmen and 1 labourer will mix and lay 1 m² in 0.2 hours:
   Cost per m²: £17.50 × 0.20   3.50
Additional labour for width <300 mm: say 25%   0.87

Cost per m²   4.85

Cost per metre for width 150 mm: £4.85 × (150/1000)   0.73
Profit and oncost 20%   0.15

Rate per metre   £0.88

Example 15.11

> Rounded angles and intersections: on cement/sand plaster; 10–100 mm radius.   m

Labour
Tradesman will form 17 m in 1 hour
Cost per metre: 1/17 hours at £6.10   £0.37
Profit and oncost 20%   0.07

Rate per metre   £0.44

Roughcast coatings

Table 15.5 shows a range of labour outputs for applying roughcast coatings. The hours are again based on a squad consisting of 2 tradesmen and 1 labourer.

Table 15.5 *Range of labour outputs for roughcast*

Description	Squad hours per 100 m²
Roughcast dry dash	19–24
Roughcast wet dash	22–27

Example 15.12

> Roughcast: first coat of cement, lime and sand 2:1:9; second coat of cement, lime and coloured sand 1:1:6; third coat of spar chips, dry dash; 19 mm work to walls on brickwork or blockwork base; >300 mm wide.   m²

Materials

Lime bagged delivered to site per tonne		£89.30
Unload and stack: 1 hour/tonne at £5.30		5.30
Slaking lime: 2 hours at £5.30		10.60
		105.20

Materials cost per m^3 = tonnes/m^3 at cost/tonne:

Cement:	1.28 at £66.80	85.50
Lime:	1.67 at £105.20	175.68
Sand:	1.28 at £7.50	9.60
Coloured sand:	1.28 at £24.00	30.72
Chips:	1.60 at £35.50	56.80

First coat 2:1:9 mix by volume:

2 m^3 of cement at £85.50	171.00
1 m^3 of lime at £175.68	175.68
9 m^3 of sand at £9.60	86.40
	433.08
Allow 20% shrinkage: therefore add 25%	108.27
Cost 12 m^3 of 2:1:9 mix	541.35
Cost 1 m^3 of 2:1:9 mix	45.11

Cost per m^2 at 10 mm thick: £45.11 × 0.01	£0.45
Wastage 5%	0.02

Second coat 1:1:6 mix by volume:

1 m^3 of cement at £85.50	85.50
1 m^3 of lime at £175.68	175.68
6 m^3 of coloured sand at £30.72	184.32
	445.50
Allow 20% shrinkage: therefore add 25%	111.38
Cost 8 m^3 of 1:1:6 mix	556.88
Cost 1 m^3 of 1:1:6 mix	69.61

Cost per m^2 at 9 mm thick: £69.61 × 0.009	0.63
Wastage 2$\frac{1}{2}$%	0.02

Third coat: 1 m^3 chips will cover 110 m^2

Cost per m^2: £56.80/110	0.52
Wastage 5%	0.03

Labour

Squad of 2 tradesmen and 1 labourer will produce 1 m^2 in 0.23 hours, including fixing and removing temporary grounds:

Cost per m^2: £17.50 × 0.23	4.03

Mixer

In this example we shall include the cost of a mixer rather than presume that its cost is covered in the preliminaries. The costs for mixing and transporting are included in the above labour costs. A mixer would normally be shared between a number of squads: in this example say 4 squads to share.

Cost of mixer per hour: say £1.70
Cost per m²: (0.23 hours at £1.70)/4   0.10
                                                                         5.80
Profit and oncost 20%   1.16
Rate per m²   £6.96

Example 15.13

> Roughcast: first coat of cement, lime and sand 2:1:9; second coat of
> cement, lime and coloured sand 1:1:6; third coat of cement and granite
> chips 1:2 wet dash; 27 mm work to walls on brickwork or blockwork
> base; >300 mm wide.   m²

Materials
Materials cost per m³ = tonnes/m³ at cost/tonne
    Cement: 1.28 at   £66.80   £85.50
    Granite chips: 1.67 at   £19.00   31.73

First coat per m² at 10 mm thick (Example 15.12) including waste   0.47
Second coat: cost per m² at 9 mm thick (Example 15.12) including waste 0.65

Third coat 1:2 mix by volume:
    1 m³ of cement costs   85.50
    2 m³ of chips costs   63.46
                                                                     148.96
Allow 20% shrinkage: therefore add 25%   37.24
Cost 3 m³ of 1:2 mix   186.20
Cost 1 m³ of 1:2 mix   62.07
Cost per m² at 8 mm thick: £62.07 × 0.008   £0.50
Wastage 5%   0.03

Labour
Squad of 2 tradesmen and 1 labourer will produce 1 m² in 0.27 hours, including fixing
and removing temporary grounds:
    Cost per m²: £17.50 × 0.27   4.73

Mixer
Cost per m²: (see Example 15.12): (0.27 hours at £1.70)/4   0.10
                                                                     6.48
Profit and oncost 20%   1.30
Rate per m²   £7.78

Beads

Table 15.6 shows a range of labour outputs for fixing beads.

Table 15.6 *Range of labour outputs for beads*

Description	Tradesman m/hour
Metal angle bead	18–20
Metal stop bead	18–20

Example 15.14

Galvanized steel angle beads: fixing with plaster dabs; reference 553. m

Materials	
Angle bead in 3 m lengths at £0.90: per metre	£0.30
Plaster dabs: per metre say	0.10
	0.40
Wastage 5%	0.02
Labour	
Tradesman will fix 20 m in 1 hour	
Cost per metre: 1/20 hours at £6.10	0.30
	0.72
Profit and oncost 20%	0.14
Rate per metre	£0.86

Metal mesh lathing

Table 15.7 shows a range of labour outputs for fixing expanded metal lath. The hours are again based on a squad consisting of 2 tradesmen and 1 labourer.

Table 15.7 *Range of labour outputs for expanded metal lath*

Description	Squad hours per 100 m^2
Expanded metal lath to walls	18–22
Expanded metal lath to ceilings	23–26

Example 15.15

Lathing: galvanized expanded metal; butt joints; fixing with galvanized nails; 6 mm mesh to walls on timber base; >300 mm wide. m^2

Materials

Cost of steel lath at £210 per 100 m²	£210.00
Take delivery and stack 100 m²: 1 hour at £5.30	5.30

Galvanized nails to fix 2500×700 mm sheet at 150 mm centres (sheets overlap, effective area 2400×600 mm):

    To timber members at 600 mm metres: $2 \times [(2400/150) + 1] = 34$

    To dwangs at 1200 mm centres: $3 \times [(600/150) + 1 - 2] = 9$

Nr nails per sheet: $34 + 9 = 43$

Nr nails per 100 m²: $43 \times 100/(2.4 \times 0.6) = 2986$

Cost per 100 m²: 2986 at 110/kg at £2.50/kg	67.86
	283.16
Wastage 10%	28.32
	311.48

Labour

Squad of 2 tradesmen and 1 labourer will fix 100 m² of lath in 20 hours:	
Cost per 100 m²: £17.50 × 20	350.00
	661.48
Profit and oncost 20%	132.30
Cost per 100 m²	793.78
Rate per m²	£7.94

Fibrous plaster work

A typical output for a tradesman fixing plaster cove is 8–12 m/hour.

Example 15.16

Fibrous plaster cove: plain; 100 mm girth.	m

Materials

Pre-formed cove in 3 m lengths at £3.90: per metre	£1.30
Cove adhesive 5 kg pack at £3.80 will fix 20 m: £3.80/20	0.19
	1.49
Wastage 5%	0.07

Labour

Tradesman will fix 10 m in 1 hour	
Cost per metre: 1/10 hour at £6.10	0.61
	2.17
Profit and oncost 20%	0.43
Rate per metre	£2.60

Ceramic and quarry tiling

Table 15.8 shows a range of labour outputs for tiling. The hours are again based on a squad consisting of 2 tradesmen and 1 labourer.

Table 15.8 *Range of labour outputs for tiling*

Description	Squad hours per 100 tiles
Ceramic wall tiles 5.5 mm thick:	
150 × 150 mm	0.38–0.43
100 × 100 mm	0.35–0.40
Quarry tiles to floors 16 mm thick:	
225 × 225 mm	1.35–1.45
150 × 150 mm	1.25–1.35
150 × 150 mm cove skirting	1.75–1.85

Example 15.17

> Ceramic tiles: plain white glazed; butt joints; fixing with adhesive; grouting horizontal and vertical joints with anti-fungal cement; 150 × 150 × 5.5 mm units to walls on plaster base; over 300 mm wide.   m²

Materials

Tiles supplied in packs of 18 tiles at £4.90 per pack	
Area covered by 100 tiles: $100 \times 0.152 \times 0.152 = 2.31$ m² (2 mm joints)	
Cost of 100 tiles: $(100/18) \times £4.90$	£27.22
Unload and stack 100 tiles (5.55 packs) in 6 minutes: $0.1 \times £5.30$	0.53
Adhesive: 0.6 litres at £3.20 per litre	1.92
Cement grout: 0.5 kg at £0.95 per kg	0.48
	30.15
Wastage 5%	1.51

Labour

Squad of 2 tradesmen and 1 labourer will fix 100 tiles in 0.40 hours:	
Cost per 100 tiles: $£17.50 \times 0.40$	7.00
	38.66
Profit and oncost 20%	7.73
Cost per 100 tiles	46.39
Rate per m²: £46.39/2.31	£20.08

Example 15.8

> Quarry tiles: butt joints; including 10 mm bed of cement/sand mortar
> 1:3; jointing and pointing with cement/sand mortar 1:3; to floors level
> or to falls less than or equal to 15 degrees; $150 \times 150 \times 16$ mm units;
> plain; symmetrical pattern; over 300 mm wide $\qquad$ m^2

Materials

Tiles supplied in packs of 10 tiles at £5 per pack
Area covered by 100 tiles: $100 \times 0.154 \times 0.154 = 2.37$ m^2 (4 mm joints)

Cost of 100 tiles: $(100/10) \times £5$	£50.00
Unload and stack 100 tiles (10 packs) in 15 minutes: $0.25 \times £5.30$	1.32
Mortar bed for 100 tiles at 10 mm thick: 0.0237 m^3 at £35.72/m^3	0.85
Mortar grout: say 0.005 m^3 at £35.72/m^3	0.18
	52.35
Wastage 5%	2.62

Labour

Squad of 2 tradesmen and 1 labourer will fix 100 tiles in 1.25 hours:	
Cost per 100 tiles: £17.50 × 1.25	21.88
	76.85
Profit and oncost 20%	15.37
Cost per 100 tiles	92.22
Rate per m^2: £89.02/2.25	£38.91

Example 15.19

> Quarry tile skirting; butt joints; including 10 mm bed of cement/sand
> mortar 1:3; jointing and pointing with cement/sand mortar 1:3; to
> floors; 150 mm high; $150 \times 150 \times 16$ mm units; round top, cove base. $\qquad$ m

Materials

Skirting tiles cost £0.70 each
Length covered by 100 tiles: $100 \times 0.154 = 15.4$ m

Cost of 100 tiles: $100 \times £0.70$	£70.00
Unload and stack 100 tiles in 15 minutes: $0.25 \times £5.30$	1.32
Mortar bed for 100 tiles at 10 mm thick: 0.0237 m^3 at £35.72/m^3	0.85
Mortar grout: say 0.005 m^3 at £35.72/m^3	0.18
	72.35
Wastage $2\frac{1}{2}$%	1.81

Labour

Squad of 2 tradesmen and 1 labourer will fix 100 tiles in 1.75 hours:	
Cost per 100 tiles: £17.50 × 1.75	30.63
	104.79
Profit and oncost 20%	20.96
Cost per 100 tiles	125.75
Rate per metre: £125.75/15.4 m	£8.17

16

Glazing

SMM7 Sections L40

Introduction

The use of glass in the building industry has increased greatly over the last few decades. The versatility of flat glass – the generic term used to describe unbowed glass used in the glazing of windows and doors – has enabled it to be used in many ways which ultimately enhance the quality of life. Glass has been adapted to reduce both heat loss and noise with double and triple glazed units. Laminated and toughened glass offer greater safety than ordinary sheet and float glass; laminated glass gives the added advantage of increased security protection. Wired glass prevents the spread of flames, while solar control glasses (specially coated and tinted) help reduce the effects of radiant heat from the sun. Patterned and obscured glass afford a greater degree of privacy together with a decorative quality.

There are also numerous materials for the fixing of glass to frames, from traditional linseed oil and whiting putty, timber beads screwed or nailed to frames, to modern sealants, non-setting compounds, mastic tapes and so on. We shall however be concentrating on the basic methods by which glazing is fixed to frames.

Estimating for glazing

In order to estimate the costs of a particular item of work accurately, it is necessary to have details of the following:

- Glass type, quality, thickness and normal maximum sizes of sheets
- Type of surround to which glass is to be fixed, e.g. wood, aluminium, uPVC
- Glazing compounds, beads etc. and their suitability for work in connection with a particular surround.

Wastage varies dependent on whether glass is bought precut to the required size, or in large sheets which are cut to size on site or in the contractor's yard. Typical wastage figures are as follows:

Precut glass	5%
Uncut glass	10%
Glazing compounds and sundries	10%

Under the rules laid down in SMM7 for glazing, standard plain glass is grouped into two main categories of pane area: less than 0.15 m^2, and from 0.15 to 4 m^2. Non-standard plain glass, that is any glass (other than special glass) which is drilled, brilliant cut or bent, or is 10 mm thick or more, or is in panes greater than 4 m^2, gives rise to another category of pane area: more than 4 m^2.

The variance in pane sizes within the category 0.15 to 4 m² can be great. Therefore the estimator will assess the average pane size and allow outputs accordingly. Although the estimator may choose to determine his own categories and relate the relevant outputs etc. to those categories, it is convenient in our case to adopt the categories used in SMM6.

Table 16.1 shows typical requirements for glazing compound. Table 16.2 shows outputs for a tradesman for cutting and glazing.

Types of glass

Clear sheet glass

Sheet glass, because of the nature of its manufacture, never has two surfaces perfectly parallel or flat; it thus distorts transmitted and reflected light. This glass is manufactured in thicknesses of

Table 16.1 *Approximate quantities of glazing compound for back and front fixing*

Pane sizes: adapted SMM7 categories (m²)	Wood surrounds* (kg/m²)	Metal surrounds* (kg/m²)
≤0.15	3.00	4.00
>0.15 to 0.5	2.00	2.66
>0.5 to 1	1.00	1.33
>1	0.75	1.00

*For back putty only, 33.33 per cent of above weights should be used.

Table 16.2 *Tradesman's average outputs for cutting and glazing (full back and front fixing) (hours/m²)*

Pane size (m²)	Clear sheet ≤5 mm thick Cut	Glazing Wood	Metal	Obscured, patterned or float ≤5 mm thick Cut	Glazing Wood	Metal	Clear sheet, roughcast, polished plate >5 mm but ≤6 mm thick Cut	Glazing Wood	Metal	Polished plate, float 10–12 mm thick and 6 mm wired glass Cut	Glazing Wood	Metal
≤0.15	0.45	1.10	1.20	0.55	1.20	1.35	0.60	1.25	1.40	0.70	1.50	1.65
>0.15 to 0.5	0.35	0.80	0.90	0.45	0.90	1.00	0.50	0.95	1.05	0.55	1.20	1.35
>0.5 to 1	0.25	0.60	0.65	0.30	0.65	0.70	0.35	0.75	0.85	0.45	1.00	1.10
>1	0.20	0.45	0.50	0.25	0.50	0.55	0.30	0.60	0.65	0.35	0.80	0.90

Glazing with bradded wood beads — deduct 10%
Glazing with screw beads to wood — add 15%
Glazing with metal screw beads to metal — add 20%

Hacking out glass and preparing rebates to receive new glass — 0.2 hours/m
Curved cutting, depending on type of glass and thickness — 1.5–2.5 hours/m

2, 3, 4, 5 and 6 mm (the 2 mm thickness is not recommended for general glazing). Clear sheet glass has generally been superseded by float glass.

The quality of the glass used depends on the class of work desired. BS 952 recommends three qualities:

Ordinary glazing quality	OQ
Selected glazing quality	SQ
Special selected quality	SSQ

Float glass

This is a truly flat glass, giving undistorted vision. It is formed by floating a continuous ribbon of molten glass on to liquid metal at a controlled rate and temperature. This glass is manufactured in general glazing quality and selected quality in thicknesses of 3, 4, 5, 6, 10 and 12 mm.

Polished plate glass

Polished plate glass has both surfaces ground and polished and in parallel, giving undistorted vision. It has generally been superseded by float glass, which has the same qualities at reduced cost. This glass is manufactured in thicknesses of 5, 6, 10 and 12 mm, although it can be manufactured up to 38 mm thick.

Patterned glass

Patterned glass is a rolled glass. During manufacture a pattern is formed on one face of the glass while the other face is left flat. The pattern makes the glass partially or totally obscure. This glass is manufactured in thicknesses of 3 and 5 mm.

Roughcast glass

Roughcast glass is a translucent glass which has a textured surface introduced by the rollers during manufacture. It is manufactured in thicknesses of 5, 6 and 10 mm.

Wired glass

Wired glass has wire mesh incorporated during the rolling process. Wired glass may have a cast finish or be subsequently polished. It is manufactured in thicknesses of 6 mm polished or 7 mm cast.

Laminated glass

This glass consists of a sandwich construction of two or more sheets of ordinary glass, between which are bonded interlayer(s) made of resilient plastic material. It is manufactured in thicknesses of 4.4, 5.4, 6.1, 6.4, 6.8 and 8.8 mm.

Toughened glass

Toughened glass is formed by heat treating clear, patterned and other types of glass (wired glass cannot be treated in this manner). It is manufactured in thicknesses of 4, 5, 6 and 10 mm.

Example 16.1

4 mm OQ clear sheet glass to 20 mm wood rebates with putty; panes 0.15 to 4 m² (average pane sizes within category 0.15–0.5 m²).	m²

Materials

Glass delivered to site per m²	£9.73
Putty: 2 kg at £9.15 per 25 kg	0.73
Glazing sprigs: say	0.03
	10.49
Allow waste (precut glass) 5%	0.52
	11.01

Labour

Tradesman 0.8 hours at £6.10	4.88
	15.89
Profit and oncost 20%	3.18
Rate per m²	£19.07

Example 16.2

4 mm OQ clear sheet glass to 20 mm metal rebates with metal; glazing compound; not exceeding 0.15 m². 30 nr.	m²

Materials

Glass delivered to site per m²	£5.27
Glazing compound: 4 kg at £9.95 per 25 kg	1.59
Steel sash clips: say	0.12
	6.98
Allow waste (uncut glass) 10%	0.70
	7.68

Labour

Tradesman 1.65 hours at £6.10	10.06
	17.74
Profit and oncost 20%	3.55
Rate per m²	£21.29

Example 16.3

4 mm patterned glass to 20 mm wood rebates with putty; panes 0.15 to 4 m² (average pane sizes within category 0.15–0.5 m²).	m²

Materials

Glass delivered to site per m²	£5.92
Putty: 2 kg at £9.15 per 25 kg	0.73
Glazing sprigs: say	0.03
	6.68
Allow waste (uncut glass) 10%	0.67
	7.35

Labour

Tradesman 1.35 hours at £6.10	8.24
	15.59
Profit and oncost 20%	3.12
Rate per m²	£18.70

Example 16.4

> 4 mm float glass to 20 mm wood rebates with bradded wood beads; panes 0.15 to 4 m² (average pane sizes within category 0.5–1 m²).   m²

Materials

Glass delivered to site per m²	£5.27
Putty: 0.33 kg at £9.15 per 25 kg	0.12
Brads (beads supplied by joiner, left loose): say	0.06
	5.45
Allow waste (uncut glass) 10%	0.55
	6.00

Labour

Tradesman 1.35 hours at £6.10	8.23
	14.23
Profit and oncost 20%	2.85
Rate per m²	£17.08

Sealed multiple glazed units

Double or multi glazed units are factory made using two or more panes. The panes are not necessarily of the same type; a selection can be made from the flat glasses, such as float, patterned or wired. The panes are separated by special profiled spacers, ranging from 6 to 20 mm wide. The perimeter is sealed using an impervious, moisture resistant, polysulphide material. The enclosed space is filled with dry air or an inert gas. Double or multi glazed units are purpose made or in stock sizes for standard windows.

Example 16.5

Pilkington Insulight double glazing units; 2 nr 4 mm float glass with 6 mm air space; to wood rebates with general purpose butyl strip compression joint sealant and wood beads, screwed; 1830 × 1830 mm.	Nr

Materials

Unit size 1830 × 1830 mm delivered to site	£131.45
Butyl strip 2 × 4 × 1830 = 14.64 mm at £6.50 per 10 m	9.52
	140.97
Allow waste 5%	7.05
	148.02

Labour

Tradesman 3 hours at £6.10	18.30
	£166.32
Profit and oncost 20%	33.26
Each	£199.58

Roof and wall glazing

Patent glazing systems

Patent glazing systems form a non-load-bearing glazed vertical or pitched external envelope to buildings. They consist generally of a framework of extruded aluminium alloy or PVC clad glazing bars, which supports single glazing panes or double glazing units. Each system incorporates its own standard or purpose made flashing, trims, gutters, expansion joints etc. No pricing examples are given of this form of construction, as individual project designs would be priced by a specialist company and included in a building contract as a nominated subcontract. Of the three types of roof and wall glazing described in this section, patent glazing is normally the cheapest.

Curtain walling systems

Curtain walling systems again provide a glazed external envelope to buildings in the form of a lightweight cladding attached to the structural framework of a building. The basic concept of curtain walling is a series of vertical mullions fixed to the structural frame by connectors which support the cladding's own deadweight and the imposed wind loads. Horizontal transoms form the connecting framework, to which are fixed glass panes or units and infill panels. As for patent glazing, no pricing examples are given of this form of construction, as individual project designs would be priced by a specialist company and included in a building contract as a nominated subcontract.

Structural glazing systems

Structural glazing is a form of curtain walling construction incorporating specially developed high performance silicone materials, by which glass panes or units are adhered to non-structural mullions and transoms. These are in turn fixed to the structural framework in a manner similar to curtain walling. The effect of a glass-only facade can be achieved using this type of system. Again no examples of price build-up are given owing to the specialist nature of the construction.

17

Decorative papers and painting
SMM7 Sections M52, M60

Painting and clear finishing

In the determination of the covering capacities of materials and the labour outputs for decoration work, the estimator is faced with more variables than with most other trades. For example, walls may vary greatly in porosity or texture. Therefore average coverages and outputs are given in Tables 17.1 and 17.2.

Under the coverage rules of SMM7, the preparation of surfaces and the rubbing down between coats with glass-, emery- or sand-paper are deemed to be included. Any other preparatory work or treatment work required, such as wire brushing, should be detailed in the

Table 17.1 *Covering capacities per 100 m² for decoration work*

Material	Surface	Quantity required to cover 100 m² (litres except as given)
Primer	Wood	8.0
	Metal	6.5
	Plaster	8.3
	Brick	12.5
	Concrete	9.0
Undercoat	Wood	7.2
	Metal	7.2
	Plaster	7.2
	Brick	7.2
	Concrete	7.2
Finishing coat	Wood	8.3
	Metal	8.3
	Plaster	8.3
	Brick	8.3
	Concrete	8.3
Emulsion	Plaster	8.3
Varnish size	Wood	7.0
Varnish	Wood	5.6
Knotting	Wood	0.75
Stopping (putty)	Wood	2.5 kg
Sandpaper	Preparing	8 sheets
Masonry sealer	Roughcast	25
Masonry paint	Roughcast	29

Table 17.2 *Labour outputs for decoration work*

Operation and surface	Tradesman's hours per 100 m^2
Washing, rubbing down existing gloss	7.0
Rubbing down wood	4.5
Wire brushing:	
Roughcast	7.0
Concrete	3.0
Metal	5.0
Knotting and stopping	6.0
Priming:	
Wood	5.8
Metal	5.0
Plaster	6.0
Brick	7.5
Concrete	5.8
Undercoat:	
Wood	5.0
Metal	4.7
Plaster	5.8
Brick	7.0
Concrete	5.7
Finishing coat:	
Wood	4.5
Metal	4.5
Plaster	4.5
Brick	6.7
Concrete	4.5
Size to wood	6.3
Varnish to sized wood	5.3
Emulsion:	
Unprimed plaster	6.3
Plaster	5.8
Paper	5.8
Concrete	6.0
Masonry paint sealer:	
Roughcast	7.3
Brick	7.0
Concrete	6.0
Masonry paint:	
Roughcast	7.0
Brick	6.8
Concrete	5.8

specification or in bill items. A minimum preparation for sanding down is therefore taken in all items.

The labour outputs and covering capacities given in the tables are for general surfaces over 300 mm girth. Therefore the figures have to be adjusted, for example, where working to narrower girths or to window frames. Table 17.3 lists percentage adjustments for labour and materials. The covering capacities decrease, resulting in an increase in the costs of materials. Similarly, because of the increased difficulty of the work, the labour outputs decreases and thus the labour costs increase.

It is generally recognized that overheads incurred in the running of a painting and decorating company are not as great as those of a general builder. This can be attributed to a number of factors, namely fewer staff to administer, smaller premises to rent or buy, less investment in plant, and so on. In the examples that follow we allow 15 per cent for profit and oncost.

Table 17.3   *Labour and material cost increases on involved decoration work*

Description	Percentage increase in labour costs	Percentage increase in material costs
Isolated surfaces, girth $\leqslant 300$ mm	30	10
Glazed windows and screens:		
Panes $\leqslant 0.1$ m^2	50	20
Panes 0.1 to 0.5 m^2	25	10
Panes 0.5 to 1 m^2	10	nil
Panes > 1 m^2	nil	less 10
Radiators, panel type:		
Over 300 mm girth	20	10
Isolated surfaces $\leqslant 300$ mm girth	30	15
Railings, fences and gates, plain open type:		
Over 300 mm girth	10	5
Isolated surfaces $\leqslant 300$ mm girth	30	10
Gutters, eaves:		
Over 300 mm girth	10	5
Isolated surfaces $\leqslant 300$ mm girth	30	10
Services, pipes:		
Over 300 mm girth	10	5
Isolated surfaces $\leqslant 300$ mm girth	40	15

Example 17.1

> Preparing, knotting, stopping, one coat primer, two coats undercoat,
> one coat gloss finish to general surfaces of wood; over 300 mm girth.   m^2

The cost is built up for 100 m^2

Materials
Sandpaper preparing	8 sheets at £0.15	£1.20
Putty stopping	2.5 kg at £1.07	2.67

Knotting	0.75 litres at £6.60	4.95
Primer	8 litres at £6.99	55.92
Undercoat (two coats)	14.4 litres at £4.15	59.76
Finish	8.3 litres at £4.15	34.44
		158.95

Labour

Preparing	4.5	
Stopping and knotting	6	
Primer	5.8	
Undercoat (two coats)	10	
Finish	4.5	
	30.8 hours at £6.10	187.88
		346.83

Brushes

A brush has an expected life of 100 hours. Say painter uses brush 66.66 per cent of working hours. Then:

Painter hours a brush will last: $100 \times (100/66.66) = 150$

Brush costs say £5.50: cost per hour £5.50/150 = 0.0367

Brush cost: 30.8 hours at £0.0367	1.13
	347.96
Profit and oncost 15%	52.19
Cost per 100 m²	400.15
Rate per m²	£4.00

Example 17.2

> Preparing, knotting, stopping, one coat primer, two coats undercoat, one coat gloss finish to general surfaces of wood; isolated surfaces <300 mm girth.   m

The cost is built up for 100 m².

Materials as Example 17.1	£158.95
Decreased covering capacity: add 10%	15.89
Labour as Example 17.1	187.88
Additional labour required: add 30%	56.36
Brushes as Example 17.1	1.13
Additional for brushes: add 30%	0.34
Cost per 100 m²	420.56
Cost per 1 m²	£4.21
Cost per metre for say 150 mm: £4.21 × (150/1000)	0.63
Profit and oncost 15%	0.09
Rate per metre	£0.73

Example 17.3

> Two coats emulsion paint to general surfaces of plaster walls; over 300 mm girth.   m²

The cost is built up for 100 m².

Materials
Emulsion paint (two coats): 16.6 litres at £3.30   £54.78

Labour
First coat   6.3
Second coat   5.8
               12.1 hours at £6.10   73.81
                                     128.59

Brushes
Brush cost 12.1 hours at £0.0367   0.44
                                   129.03
Profit and oncost 15%   19.36
Cost per 100 m²   148.39

Rate per m²   £1.48

Example 17.4

> Preparing, one coat primer, two coats undercoat, one coat gloss finish to services; pipes; isolated surfaced < 300 mm girth.   m

The cost is built up for 100 m².

Materials as Example 17.1		£158.95		
Deduct knotting and stopping	7.62			
Deduct 1.5 litre priming at £6.99	10.49	18.11	140.84	
Decreased covering capacity: add 15%			21.13	161.97
Labour as Example 17.1		187.88		
Deduct knotting and stopping: 6 hours at £6.10	36.60			
Deduct priming 0.8 hrs at £6.10	4.88	41.48	146.40	
Additional labour required: add 40%			58.56	204.96
Brushes as Example 17.1		1.13		
Deduct knotting and stopping: 6 hours at £0.0367	0.22			
Deduct priming 0.8 hrs at £0.0367	0.03	0.25	0.88	
Additional for brushes: add 40%			0.35	1.23
Cost per 100 m²				368.16

Cost per 1 m²	3.68
Cost per metre for say 150 mm: £3.68 × (150/1000)	0.55
Profit and oncost 15%	0.08
Rate per metre	£0.63

Example 17.5

> Two coats clear varnish to general surfaces of wood doors; over
> 300 mm girth.　　　　　　　　　　　　　　　　　　　　　　　m²

The cost is built up for 100 m².

Materials

Varnish size:	7 litres at £6.50	£45.50
Varnish (two coats):	11.2 litres at £3.30	36.96

Labour

Size coat	6.3	
Varnish coats	10.6	
	16.9 hours at £6.10	103.09
		185.55

Brushes

Brush cost: 16.9 hours at £0.0367	0.62
	186.17
Profit and oncost 15%	27.93
Cost per 100 m²	214.10
Rate per m²	£2.14

Example 17.6

> Wire brushing surfaces of existing roughcast walls; one coat masonry
> paint sealer; two coats masonry paint; to general surfaces over 300 mm
> girth.　　　　　　　　　　　　　　　　　　　　　　　　　　m²

The cost is built up for 100 m².

Materials

Wire brushes: 2 at £4.50	£9.00
Masonry sealer: 25 litres at £5.50	137.50
Masonry paint (two coats): 58 litres at £5.80	336.40

Labour

Wire brushing	7	
Sealer coat	7.3	
Masonry paint coats	14	
	28.3 hours at £6.10	172.63
		655.53

Brushes

Brush cost: 28.3 hours at £0.0367	1.04
	656.57
Profit and oncost 15%	98.49
Cost per 100 m²	755.06
Rate per m²	£7.55

Decorative papers

Decorative papers are coverings applied to internal walls or ceilings with glue or paste. There are plain and decorative papers, and fabrics and plastics on paper backings. Papers are sold in rolls or pieces to a standard size of 10 m length by 530 mm width. There are instances where paper sizes may differ from this standard, which would have to be accounted for in the development of rates.

A decision on the exact pattern of paper to be used may not have been made at the time of producing bills of quantities. However, if the quality or price range is known then a prime cost sum can be calculated. The prime cost sum will include the supply and delivery of the required paper to the site.

Table 17.4 shows wastage values for decorative papers. Table 17.5 gives labour outputs for straight cutting, pasting and hanging papers.

Table 17.4   *Average wastages for decorative paper*

Description	Percentage waste
Lining paper to walls	7.5
ceilings	5.0
Plain paper to walls	10.0
ceilings	7.5
Pattern paper to walls	15.0
ceilings	12.5
Borders and strips	7.5

Table 17.5   *Labour outputs for straight cutting, pasting and hanging decorative papers*

Description	Tradesman hours per roll
Strip old paper from walls	0.53
ceilings	0.64
Clean down, stop walls	0.27
ceilings	0.32
Glue size wall or ceilings	0.37
Light papers to walls	1.00
ceilings	0.80
Heavy papers to walls	1.33
ceilings	1.00
Borders 40 mm wide	20 m/hour

Example 17.7

Stripping existing decorative paper, stopping, glue size and hang heavy patterned decorative papers (PC sum £5.50/roll) to plaster surfaces of walls. m^2

The cost is built up for one roll.

Materials

Sand paper preparing	0.5 at £0.15	£0.07
Stopping	0.1 kg at £0.80	0.08
Glue size	1 litres at £0.15	0.15
Decorative paper	1 roll at £5.50	5.50

Labour

Strip old paper	0.53	
Prepare and stop	0.27	
Glue size	0.37	
Hang paper	1.33	
	2.50 hours at £6.10	15.25
		21.05
Profit and oncost 15%		3.16
Cost to hang 1 roll (5.3 m^2)		24.21
Rate per m^2		£4.57

18

Drainage

SMM7 Sections R12, R13

Excavating drainage trenches

The SMM7 unit of measurement for excavating drainage trenches is the metre, unlike groundworks (see Chapter 5) where the unit of measure is the cubic metre. The costs for excavation of drainage trenches, disposal of surplus excavated material, and earthwork support can be developed using the same labour constants as those in Chapter 5. However, the m^3 rates for excavation have to be adjusted by applying the volume of a metre run of trench to establish a metre rate.

The estimator has to cost certain items deemed to be included by the coverage rules of SMM7. These items are for earthwork support, consolidation of trench bottoms, trimming excavations, filling with and compaction of general filling materials, and disposal of surplus excavated material.

As no width of trench is required to be given in the item description, we have to establish widths suitable for the purpose of laying pipes. The following are acceptable trench widths for pipes up to 200 mm diameter:

Depth of trench (m)	$\leqslant 1$	$\leqslant 2$	$\leqslant 4$	$\leqslant 6$
Width of trench (mm)	550	600	750	900

Trench widths for pipes over 200 mm diameter can be established by adding the diameter of the pipe to the appropriate trench width above.

Hand excavation of pipe trenches

There is no longer a requirement in SMM7 to state depth ranges in items for excavation of pipe trenches; it is necessary only to give the average depth of excavation in stages of 250 mm. However, as covered in Chapter 5, the cost of excavating by hand is calculated in 1.5 m stages, that being the recognized maximum height that a man can throw a spade full of excavated material.

We shall first determine the cost per m^3 to excavate pipe trenches, fill with and compact filling materials, and dispose of surplus excavated material, for each 1.5 m stage or throw, i.e. up to 1.5 m deep, 1.5 to 3 m deep, and 3 to 4.5 m deep. The costs are then used in subsequent examples.

Depth stage up to 1.5 m

Excavate and get out	2.5
Fill, compact, dispose of surplus	1.5
Labour output	4.0 hours/m^3
Cost per m^3 4.0 at £5.30	£21.20

Depth stage 1.5 to 3 m

Excavate and get out	2.5
Extra, to throw from 3 m to 1.5 m	1.0
Fill, compact, dispose of surplus	1.5
Labour output	5.0 hours/m^3

Cost per m^3 5.0 at £5.30 · · · · £26.50

Depth stage 3 to 4.5 m

Excavate, get out	2.50
Extra to throw 3 m to 1.5 m	1.00
Extra to throw 4.5 m to 3 m	1.00
Fill, compact, dispose of surplus	1.50
Labour output	6.00 hours/m^3

Cost per m^3 6.0 at £5.30 · · · · £31.80

Example 18.1

> Excavating trenches for pipes not exceeding 200 mm nominal size; trimming excavations; earthwork support; consolidation of trench bottoms; filling with and compaction of material arising from excavations; disposal of surplus excavated material by removing from site; average 1.25 m deep.   m

Volume of trench: $1.00 \times 0.60 \times 1.25 = 0.75$ m^3 per metre of trench

Cost per m^3 for depth stage up to 1.5 m: £21.20
Cost per metre: £21.20 × 0.75 · · · · £15.90

Trimming excavations: $1.00 \times 0.60 = 0.60$ m^2 per metre of trench
Cost per metre at 0.08 labourer hours/m^2: 0.08×0.60 at £5.30 · · · · 0.25

Consolidation of trench bottoms: 0.60 m^2 per metre of trench
Cost per metre at 0.08 labourer hours/m^2: 0.08×0.60 at £5.30 · · · · 0.25

Earthwork support to sides of excavations: $2 \times 1.00 \times 1.25 = 2.50$ m^2
Cost of earthwork support (see Chapter 5): 2.50 at £6.35 per m^2 · · · · 15.88

	32.28
Profit and oncost 20%	6.46
Rate per metre	£38.74

Example 18.2

> Excavating trenches for pipes not exceeding 200 mm nominal size; trimming excavations; earthwork support; consolidation of trench bottoms; filling with and compaction of material arising from excavations; disposal of surplus excavated material by removing from site; average 2.5 m deep.   m

Volume of trench: $1.00 \times 0.75 \times 2.50 = 1.88 \, m^3$ per metre of trench
Volume of trench in stage to 1.5 m: $1.13 \, m^3$ per metre of trench
Volume of trench in stage 1.5 to 3 m: $0.75 \, m^3$ per metre of trench

Cost per m^3 for depth stage up to 1.5 m: £21.20
Cost per metre: £21.20 × 1.13   £23.96

Cost per m^3 for depth stage 1.5 to 3 m: £26.50
Cost per metre: £26.50 × 0.75   19.88

Trimming excavations: $1.00 \times 0.75 = 0.75 \, m^2$ per metre of trench
Cost per metre at 0.08 labourer hours/m^2: 0.08×0.75 at £5.30   0.32

Consolidation of trench bottoms: $0.75 \, m^2$ per metre of trench
Cost per metre at 0.08 labourer hours/m^2: 0.08×0.75 at £5.30   0.32

Earthwork support to sides of excavations: $2 \times 1.00 \times 2.50 = 5.00 \, m^2$
Cost of earthwork support: 5.00 at £7.00 per m^2 35.00
                                                                          _____
                                                                          79.48
Profit and oncost 20%   15.90
                                                                          _____
Rate per metre   £95.38

Machine excavation of pipe trenches

The costs for excavation of pipe trenches by machine need not be determined by the same depth ranges as those used in hand excavation. Machine excavation can be calculated using deeper depth ranges, so limiting the number of range costs to be calculated. The only restriction for a machine with regard to depth is the maximum reach of a particular machine; therefore the correct machine has to be chosen to cope with a required depth. Machine outputs will decrease as the excavation deepens; Example 18.3 allows a reasonable average output per hour for each range.

Example 18.3

> Excavating trenches for pipes not exceeding 200 mm nominal size; trimming excavations; earthwork support; consolidation of trench bottoms; filling with and compaction of material arising from excavations; disposal of surplus excavated material by removing from site; average 2.5 m deep.   m

Excavator cost
An excavator/loader will be used with a backactor of capacity $0.22 \, m^3$.

Hire of machine for 39 hour week: £350.00
Driver 39 hours at £5.30 + £23.01/week   229.71
Labourer 39 hours at £5.30   206.70
                                                                   _____
Excavator cost per week   786.41

Hourly cost of excavator (deduct 10% for standing time:	
39−4=36 hours): £786.41/35	22.47
Cleaning and maintaining at 0.5 hours per 8 hour day: 0.06 hours at	
£5.89*	0.35
Fuel per hour: 2.5 litres at £0.40	1.00
Other consumables, oil etc.: say	0.20
Insurance: say	0.10
Hourly cost of excavator	24.12

Outputs
Using a backactor of capacity 0.22 m^2:
8.0 m^3/hour should be excavated to a depth of 2 m
7.2 m^3/hour should be excavated from 2 to 4 m

Cost to excavate 1 m^3 up to 2 m: £24.12/8.0=£3.02
Cost to excavate 1 m^3 from 2 to 4 m: £24.12/7.2=£3.35

Volumes
Volume of trench: $1.00 \times 0.75 \times 2.50 = 1.88 \text{ m}^3$
Volume of trench up to 2 m: $1.00 \times 0.75 \times 2.00 = 1.50 \text{ m}^3$
Volume of trench from 2 to 4 m: $1.88 - 1.50 = 0.38 \text{ m}^3$

Costing

Cost to excavate trench to 2 m: $1.50 \times £3.02$	£4.53
Cost to excavate trench from 2 to 4 m: $0.38 \times £3.35$	1.27
Trimming excavations: $1.00 \times 0.75 = 0.75 \text{ m}^2$ per metre of trench	
Cost per metre at 0.08 labourer hours/m^2: 0.08×0.75 at £5.30	0.32
Consolidation of trench bottoms: 0.75 m^2 per metre of trench	
Cost per metre 0.08 labourer hours/m^2: 0.08×0.75 at £5.30	0.32
Earthwork support to sides of excavations: $2 \times 1.00 \times 2.50 = 5.00 \text{ m}^2$	
Cost of earthwork support: 5.00 at £7.00 per m^2	35.00
Fill and compact excavated material, dispose of surplus: 1.5 hours/m^3	
Cost per metre: $1.88 \times 1.5 \times £5.30$	14.95
	56.39
Profit and oncost 20%	11.28
Rate per metre	£67.67

The amount of fill must be adjusted if a bed and a surround of a particular material are specified. The quantity of fill will decrease and the quantity of material to be disposed of will increase.

* Driver – Class N – £23.01/week – 50p/hour – Assumes front loader capacity is over 0.58 m^3 ($\frac{3}{4}$ cu yard)

Pipework: vitrified clay and uPVC

The examples that follow deal with pipework that is most commonly used in drainage systems where low volumes of waste material and water have to be handled, e.g. house drainage up to its connection with the main sewer. The two main types of pipe used in this situation are vitrified clay and uPVC. These pipes are very resilient and are relatively easy to install. Cast iron pipes are very rarely used nowadays in underground drainage mainly due to the high material and installation costs.

Example 18.4

100 mm diameter vitrified clay pipes in trenches; push fit polypropylene flexible coupling joints.	m

Materials

Pipes in standard lengths 1.6 m delivered at £1.71/m: per length	£2.74
Take 1 polypropylene coupling per length	0.98
Joint lubricant: 1 kg per 50 m pipework at £1.85/kg	
Cost per joint: £1.85 × (1.6/50)	0.06
	3.78
Deduct full load discount 17.50% (including fittings and sundries)	0.66
	3.11
Allow wastage 5%	0.16
	3.27

Labour

Unload and stack, labourer 20 lengths/hour at £5.30: per length	0.27
Pipelayer and labourer lay one length every 5 minutes:	
(£5.49 + £5.30) × (5/60)	0.90
	4.45
Profit and oncost 20%	0.89
Cost per 1.6 m length	5.34
Rate per metre: £5.34/1.6	£3.34

Example 18.5

Extra over 100 mm diameter vitrified clay pipes for single branches.	Nr

Materials

Cost of branch	£4.15
Take 2 polypropylene coupling per branch	1.96
Cost of joint lubricant (Example 18.4) 2 at £0.06	0.12
	6.23
Deduct full load discount 17.50% (including fittings and sundries)	1.09
	5.14
Allow wastage 2.5%	0.13
	5.27

Labour
Unload and stack, labourer 40 branches/hour at £5.30: per branch	0.13
Pipelayer and labourer lay one branch every 15 minutes: £10.79 × (15/60)	2.70
	8.10
Deduct length of pipe occupied by branch (see Example 18.4 for cost/m):	
0.75 m at £4.45/1.60	2.09
Extra over rate for branch	6.01
Profit and oncost 20%	1.20
	£7.21

Example 18.6

Extra over 100 mm diameter vitrified clay pipes for bends.	Nr

Materials
Cost of bend	£1.97
Take 1 polypropylene coupling per bend	0.98
Cost of joint lubricant as Example 18.4	0.06
	3.01
Deduct full load discount 17.50%	0.53
	2.48
Allow wastage 2.5%	0.06
	2.54

Labour
Unload and stack, labourer 40 bends/hour at £5.30: per bend	0.13
Pipelayer and labourer lay one bend every 10 minutes: £10.79 × (10/60)	1.80
	4.47
Deduct length of pipe occupied by bend (see Example 18.4 for cost/m):	
0.3 m at (£4.45/1.6)	0.83
Extra over rate for bend	0.73
Profit and oncost 20%	3.64
	£4.37

Example 5.2

150 mm diameter vitrified clay pipes in trenches; push fit flexible socket and spigot joints.	m

Materials
Pipes in standard lengths 1.5 m delivered at £5.88/m: per length	£8.82
Joint lubricant: 1 kg per 70 m pipework at £1.85/kg	
Cost per joint: £1.85 × (1.5/70)	0.04
	8.86
Deduct full load discount 17.50%	1.55
	7.31
Allow wastage 5%	0.37
	7.68

Labour

Unload and stack, labourer 20 lengths/hour at £5.30: per length	0.27
Pipelayer and labourer lay one length every 8 minutes: $10.79 \times (8/60)$	1.44
	9.39
Profit and oncost 20%	1.88
Cost per 1.5 m length	11.27
Rate per metre: £11.27/1.5	£7.51

Example 18.8

110 mm diameter uPVC pipes in trenches; ring seal joints.		m

Materials

Pipes in standard lengths 3 m delivered: per length	£8.55
Take 1 pipe coupler per length	2.34
Joint lubricant: 1 kg per 50 m pipework at £1.85/kg	
Cost per joint: $£1.85 \times (3/50)$	0.11
	11.00
Deduct discount say 15.00%	1.65
	9.35
Allow wastage 3%	0.28
	9.63

Labour

Unload and stack, labourer 15 lengths/hour at £5.30: per length	0.35
Pipelayer and labourer lay one length every 7 minutes: $£10.79 \times (7/60)$	1.26
	11.24
Profit and oncost 20%	2.25
Cost per 3 m length	13.49
Rate per metre: £13.49/3	£4.50

Beds and surrounds

Example 18.9

Plain *in situ* concrete: grade C20P, 20 mm aggregate; beds and coverings; 400×400 mm to 100 mm diameter pipes.		m

Materials

Volume of concrete and pipe: $0.55 \times 0.40 \times 1.00 = 0.220 \, m^3$
Volume of pipe with wall thickness 10 mm: $3.14 \times 0.06 \times 0.06 \times 1.00 = 0.011 \, m^3$
Volume of concrete only: $0.220 - 0.011 = 0.209 \, m^3$

Cost 1 m³ concrete grade C20P, 20 mm aggregate, including mixing, transporting
   and placing: £44.50
Profit and oncost 20%   8.90

Rate per m³   53.40

Rate per metre length: 0.209 × £53.40   £11.16

It is taken that concrete will be poured to the sides of the trenches; therefore no formwork has been
included in this example. It should, therefore, be noted that as the minimum width of trench is
550 mm, concrete *used* will be 0.55 × 0.40 × 1.00 = 0.22 m³. The example which follows shows the cost
effect of using formwork in the above situation.

Example 18.10

> Plain *in situ* concrete: grade C20P, 20 mm aggregate; beds and
> coverings; 400 × 400 mm to 100 mm diameter pipes.   m

Materials
Volume of concrete and pipe: 0.40 × 0.40 × 1.00 = 0.160 m³
Volume of pipe with wall thickness 10 mm: 3.14 × 0.06 × 0.06 × 1.00 = 0.011 m³
Volume of concrete only: 0.160 − 0.011 = 0.149 m³

Cost 1 m³ concrete, including mixing, transporting and placing   £44.50

                                       6.63
Cost for 0.149 m³   6.63
Formwork* to 1 m length of beds and coverings.
   2 × 1.00 × 0.40 = 0.80 m² at £25.00/m²

                                   20.00

                                   26.63

Profit and oncost 20%   5.33

Rate per metre length   £31.96

* Includes erection and striking on completion.

19

Electrical work
SMM7 Sections V, P31

Introduction

An electrical system is designed by an electrical engineer to comprise a number of manufactured pieces of equipment, outlets and fittings, all connected by given sizes of electrical cable. The work on site entails taking delivery, storing, setting or supporting in place and connecting the equipment, outlets and fittings. Therefore, because electrical work on site concerns installation and not manufacture, the processes involved in estimating tend to be relatively straightforward.

Primary switchgear

The primary switchgear and distribution equipment are the first items to be connected to the electricity authority's supply. In a domestic situation this will be the consumer unit. A large manufacturing plant might have a substation comprising transformers, fuses, metering, circuit breakers etc. Example 19.1 illustrates the estimation for the installation of a cubicle switchboard to be installed in a light industrial unit.

Example 19.1

Cubicle switchboard fixed to concrete floor, complete with all labels, fuses, MCBs, time switches, contactors etc. as shown on drawing nr 8706 (E) 01.	Nr

The unit weighs 450 kg, and therefore the total cost is that involved in taking delivery and placing in position. It is assumed that the unit will be installed in a large cupboard on the ground floor, at the end of a 30 m corridor from which access may be gained to the outside.

The constants used in this example are based on the hours per tonne expended by each of the operatives required to move the weight. It is assumed that two operatives are capable of carrying out the task involved. If a 1 tonne piece of equipment were being moved, not only would the hours based upon the constants be increased, but also three operatives might be needed to move the weight. The cost therefore rises exponentially in proportion to the weight.

Cost of unit delivered to site   £4000

Cost of installation is calculated as follows:

	hours/tonne	hours
Unload from vehicle	1.3	0.6
Jack up and place carriers	0.8	0.4
Move externally 10 m	0.5 per 10 m	0.2
Move internally 30 m	0.6 per 10 m	0.8
Turn twice	1.6 per turn	1.4
Line up to final position	1.0	0.5
Jack down and remove carriers	0.8	0.4
Final positioning and bolt down	3.0	1.4
		5.7

Total time: 2 operatives × 5.7 hours = 11.4 man hours

Total installation cost: 11.4 × £7.65	87.21
Rate for switchboard	£4087.21
Profit and oncost 20%	817.44
Rate each	£4904.65

The costs involved in the preparation of the bolts are dealt with under builder's work.

Cabling

Cabling has three material components:

- The cable support, i.e. the cable tray or conduit
- The fixing of the cable tray or conduit to the structure
- The cable itself.

Within the light industrial unit of Example 19.1, the cable is carried from the switchgear fuseboard on a cable tray in the corridor and enters plastic oval conduit for the final run to termination, i.e. the switch, lighting outlet point or power outlet point.

Example 19.2

Standard duty, return edge galvanized cable tray 300 mm wide with bolted joints fixed to softwood joists (30 m).	m

Galvanized tray per metre	£11.50
Labour fixing 0.6 hours/m at £7.65	4.59
	16.09
Profit and oncost 20%	3.22
Rate per metre	£19.31

Example 19.3

> 25 mm straight PVC oval conduit fixed with clips to blockwork.
>   Lighting outlet points in ceiling void.   nr
>   One way switch points.   nr

It should be noted that although SMM7 requires the lengths of both conduit and cable to be stated, the above description is not unusual. It is necessary in these cases to determine from the drawings the likely lengths of both conduit and cable. It is assumed, in this example, that the following lengths for 10 of each fitting have been determined from the drawings:

Total length of conduit	150 m
Total length of cable	900 m

Thus the item measured is:

> 25 mm straight PVC oval conduit fixed with clips to blockwork: m

PVC oval conduit and draw rope: per metre	£1.02
Clips for conduit: 2 per metre at £3.20/100	0.06
Pins for clips: 4 per metre at £0.25/100	0.01
	1.09
Add 15% waste	0.16
	1.25
Labour fixing 0.5 hours/m at £7.65	3.83
	5.08
Profit and oncost 20%	1.02
Rate per metre	6.10

In order to price the item in the bill of quantities it is necessary to take the product of the measured length and divide by the number of fittings:

Cost for 150 m: $150 \times £6.10 = £915$
Rate per fitting: £915/20   £45.75

Example 19.4

> Wiring with 2.5 mm^2 PVC single core drawn into conduit fixed to trays
> or ceiling joists.
>   Lighting outlet points in ceiling void.   nr
>   One way switch points.   nr

As for Example 19.2, the above description becomes the item to be priced as follows:

Wiring with 2.5 mm² PVC single core drawn into conduit, fixed to trays or ceiling joists.	m

2.5 mm² PVC single core at £21 per 100 m: per metre	£0.21
Add 15% waste	0.03

Labour fixing at 7.65/hour:

Cable to tray	300 m at 2 hours per 100 m	6.0
Cable drawn	450 m at 3 hours per 100 m	13.5
Cable to joists	150 m at 2 hours per 100 m	3.0
Total for 900 m		22.5 hours

Time for 100 m: 22.5/9 = 2.5 hours	
Labour cost per metre: 2.5 × £7.65/100	0.19
	0.43
Profit and oncost 20%	0.09
Rate per metre	0.52

As in Example 19.3, in order to price the item in the bill of quantities it is necessary to take the product of the measured length and divide by the number of fittings:

Cost for 900 m: 900 × 0.52 = £468	
Rate per fitting: £468/20	£23.40

The installation of electrical fittings is a two-stage process involving a first and second fix. During the first fix the conduits, cables and back boxes for fittings are installed. This may involve preparatory work by the builder, which is described later. Following the finishing trades (plasterer, tiler etc.), the electrician will return for the second fix to connect the wiring to the fitting and fix the fitting to the back box.

Example 19.5

13 amp, 1 gang, switched socket outlet to BS 1363:1967 with white plastic coverplate and galvanized steel box to BS 4662: 1970 for flush mounting plugged and screwed to brickwork	Nr

Materials

13 amp, 1 gang, switched socket outlet and box	£3.50
Plugs and screws: say	0.02
	3.52
Add 5% waste	0.18

Labour

1st fix: fixing box, 12 min at £7.65/hour	1.53
2nd fix: wiring socket and screwing coverplate to box, 15 min at £7.65/hour	1.91
	7.14
Profit and oncost 20%	1.43
	£8.57

Builder's work

This work is covered by Section P31 of SMM7. It comprises preparatory work, e.g. cutting chases in brick walls and forming holes, and is carried out by the builder for the electrician. The work will be undertaken by a tradesman or a labourer depending upon its complexity.

Example 19.6

Cutting or forming chase in brickwork for concealed single 25 mm straight PVC oval conduit	m

Labourer cuts 25 × 25 mm chase in brickwork with mechanical hammer at 0.125 hours/m at £5.30	£0.66
Profit and oncost 20%	0.13
Rate per metre	£0.79

Example 19.7

Cutting or forming sinking in brickwork for concealed 80 × 80 × 35 mm deep galvanized steel box for socket outlet point.	Nr

Labourer cuts 80 × 80 × 35 mm sinking in 10 min at £5.30/hour	£0.88
Profit and oncost 20%	0.18
Rate per sinking	£1.06

20

Analogous rates

Introduction

Whether you are a builder or a surveyor, you will some day be confronted with the need to produce a rate for work carried out on site which has not been priced in a bill of quantities or a schedule of rates. Such a rate is termed an 'analogous rate' or a 'pro rata rate'. While not decrying the term 'pro rata', we note that it means 'proportionally'. The purpose of this chapter is to show that not all elements of a bill rate are proportioned directly, so the preferred term is 'analogous rate'.

This chapter deals with the general principles involved in the analysis of existing rates and the build-up of analogous rates for several trades or work sections. The builder should have ready access to his estimator's original calculations and should have no difficulty working out what the new rate should be – from his point of view. This will not generally coincide with the view held by the client's quantity surveyor who, having no access to these original figures, will have to use the techniques described in this chapter to assess the validity of the builder's claim. This chapter has been written with the novice quantity surveyor in mind.

It is usual to require an analogous rate for work caused by something like a site instruction which has varied the original construction. This new rate therefore has to be worked out on the same basis as the rates in the bill or schedule. This means that certain elements in the rate will be the same as the original rates and others will differ.

The reason we wish to use a similar basis to the original rates for the analogous rates is simply that when the estimator priced the original bill he allocated the costs of many of the fixed items in his own particular way, and we must follow the same pattern or our new rates will be distorted. The quantity surveyor assessing new rates must be aware of the types of thing this involves. For example, the bill items for say external renderings may include only a portion of the total cost of scaffolding, or no allowance at all for scaffolding. In the first case the estimator may have included all of bringing scaffolding to site, erection, and taking down and removal in the bricklaying rates. Thus scaffolding for the roughcaster is only on the basis of 'cost while it is left a week or two longer'. In the second instance, the total cost of scaffolding is placed in the preliminaries and is not apportioned over the bill rates for measured work.

Elements of analogous rates

Generally the elements making up a rate for measured work are as follows:

- Materials
- Labour
- Plant
- Oncost
- Profit.

For each of these elements there are a number of factors which determine the level of cost which the estimator originally included in the bill rates; these must be analysed.

Material

The factors here are

● Quantity discounts
● Bulk buying or ordering
● Advance purchase.

The first two may seem much the same, but there are subtle differences! Put yourself in the builder's place. Order 1 tonne of OP cement and you will be quoted a price. Order 100 tonnes and the price per tonne will be a lot less. The same holds true for bulk buying or ordering, but in this case the builder places an order for 100 tonnes of OP cement in advance of getting the firm order for the work. He then 'draws' the OP cement in smaller quantities as he requires it. Not necessarily such a good price will be quoted.

Advance purchase is generally made when the builder has foreknowledge of price increases in basic materials which he knows will be utilized fairly soon. 'Fairly soon' is important because he is tying up capital or extending borrowing in the hope of making a larger profit in the reselling of the material. The additional profit must be greater than the cost of borrowing or the loss of interest on capital invested.

Labour

Particular considerations here are

● Outputs
● Bonus
● Weather
● Site management and organization.

These have all been discussed in Chapter 2, so there is no need to repeat the arguments here.

Plant

Here the estimator must consider

● Availability
● Suitability
● Level of maintenance
● Level of skill in use.

Oncost

Oncosts include head office overheads and site overheads.

Profit

The level of return required must be established.

Procedure

The procedure generally adopted to make up a rate analogous to a bill rate is as follows:

1   Establish the percentage for oncost and profit and reduce the original bill rate to net cost.
2   Isolate unknown factors by calculating the known part and deducting this from the net cost. Generally this means calculating the cost of materials, leaving labour and plant costs as the unknowns.
3   Of labour and plant costs, the easier to calculate will be the plant cost.
4   Then adjust the labour cost.
5   Work out the new plant cost.
6   Add materials costs.
7   Add oncost and profit.

Before launching into an expansion of these ideas it must be stressed that the new work for which the analogous rate is required must be comparable with the bill item which will form the basis for the calculation. It is no use trying to build up a new rate for machine excavation for a pipe trench using a rate for hand excavation of manholes. This is an extreme example, but serves as a warning that all factors have to be considered. Where, when and how is the new work to be carried out? If the answers match up with the bill item used, then go ahead.

Depending on the variation incorporated in the new item, one or all of the elements will change. For example, suppose the bill has an item for a half brick wall built in Blogg's heather facing brick, and a site instruction is issued to use Blogg's wire cut rustic buff facing brick at an increased cost of £50 per thousand. Obviously if the bricks are the same size, use the same mortar and have the same weight, then the only difference between the original and the analogous rate is the difference in cost for the bricks and the adjustment of oncost and profit on that difference. This is a fairly common problem, and will be treated more seriously as an example later in the chapter.

Assessment of oncost and profit

If we assume that the original estimator was reasonably consistent, then we can further assume that the percentages added for oncost and profit were the same for all major items.

In that case we must examine the bill and extract the rates for the principal items which carry the major part of the cost for the particular trade or work section with which we are dealing. We now carry out the process of building up our own net costs for these items. The analysis of the rates described in this step has to imitate as closely as possible the original estimator's mental process. This analysis must obviously include the rate(s) which we wish to change or to which our new rate(s) will be analogous. We subtract our net cost from the bill rate. The difference is expressed as a percentage of our net rate, and the percentage is noted. Once we have a number of rates analysed in this way we should see a pattern emerging in the percentages. We can adjust our assumptions on labour costs, materials costs and plant costs in the light of the differences in these percentages, until we are satisfied that we have duplicated as nearly as possible what the original estimator did when he priced the bill or schedule. We should now have isolated oncost and profit, although labour, plant and material will still be only notional, that is, our guestimate. However, we should be in a position to reduce our bill rate to a net cost.

Isolation of unknown factors

At this stage we generally wish to find the cost of the materials used in the bill rate, simply because it is the information most readily available. Manufacturers and suppliers can be consulted at this stage as well as price books and price listings in the technical press. One must be aware that prices quoted may or may not include the cost of transport, loading and unloading, site handling and the possibility of bulk discounts. One must also be aware that the materials cost in the bill rate is the cost to the contractor *when the work was carried out*. This may seem obvious but needs emphasizing. The analogous rate may be calculated anything up to a year or more after the contract has been completed. Price lists or quotes at that date are no help in analysing a rate for work carried out so much earlier.

Deducting the materials cost from the net rate leaves us with the two unknowns of plant and labour. Of these, labour is the more problematical. We have the difficulty of knowing the level of wages and bonus allowed for; it is very difficult to duplicate an estimator's figures in respect of hourly labour costs and outputs. One can only guess at things like squad sizes, allowances for guaranteed week, overtime factors and travelling time. So the one to tackle is the plant figure.

Calculation of plant cost

The quantity surveyor will have some idea of the make-up of the contractor's organization; for example, the contractor may own some basic machinery but will hire large or specialized items. Plant costs as well as outputs can be obtained readily from plant hirers, and one can then assess the proportion of the plant cost in the bill rate fairly easily. Subtraction of this last sum gives the labour cost.

Adjustment of labour cost

This step marks the beginning of the synthesis of the new rate(s).

In some instances the labour cost can be directly proportioned to the cost in the bill rate. If the labour will vary significantly, it may be necessary to calculate a labour cost from new output figures, squad sizes and all-in hourly rates. Look at the examples for different ways to carry this out. To the adjusted labour cost is added the new plant cost.

Addition of new plant cost

Previous work will leave the quantity surveyor with lots of information on plant costs which can now be used to give a new plant cost. It may not always be necessary to change the plant cost at all, as illustrated in some examples later in the chapter.

Addition of new materials cost

Prices for new materials are added at this point. The sources are the same as for the analysis of the bill rates.

Addition of oncost and profit

Adding oncost and profit completes the analogous rate. However, the budding quantity surveyor should not feel that this is the end of the matter. Some hard bargaining might well follow when this new rate is proudly displayed to the contractor and his surveyor. Remember, they have all the original information at their fingertips.

Interpretation of bills and schedules

We are all aware that the construction industry is never as simple as we have just described – and never forgiving of foolish generalizations. The estimator may not have put all the oncost in the item rates; the profit may all be in the percentage added on at the end of the tender; the plant and scaffold may all be in the preliminary bill; and so on. The earlier chapters on preliminaries, tendering and so on should help you look at priced bills with the right ideas in mind. You should examine bills with the object of spotting front loading, net item rates etc., and then you must be able to adjust the steps above accordingly.

Of course, there are other clues available to you. The fact that as a quantity surveyor you visit the site to carry out regular interim valuations means that you actually see what is going on; for example, you should note what general plant is kept on site. You can obtain much information from a good clerk of works. Look at the clerk of works log book and see which and what size of squads were or are on site. The clerk of works can tell you if any special provisions had to be made for some parts of the work.

Many contracts now require that the contractor prepares a *modus operandi* or plan of work: where he will use certain types of plant, the order in which work will be done, squad sizes etc. The contractor may also be expected to provide critical path networks and keep these up to date on a weekly basis. These can identify, to the knowledgeable, where the contractor has used plant, varied squad sizes etc. Earlier interim valuations with their list of materials on site should alert you to the size of loads being brought in and, for example, whether materials are being bought in bulk at a discount or favourable price.

All the clues are there; they require first to be seen, and second to be correctly interpreted.

Before continuing with a number of examples of analogous rates calculations, we will assume that analysis of principal rates has shown that the level of profit and oncost is 20%.

Example 20.1

Bill rates: machine excavation

Excavate foundation trench not exceeding 2 m deep.	m³	£4.04
Excavate foundation trench not exceeding 4 m deep.	m³	£4.51
Basement excavation not exceeding 2 m deep.	m³	£1.44

Variation account: machine excavation

Basement excavation not exceeding 4 m deep.	m³	£1.61

It is assumed that the starting level of all excavations is within 0.25 m of existing ground level. The rate of £1.61 is arrived at by direct proportion. So here we have a true pro-rata rate. The proportional increase in the basement excavation from one depth stage to another is the same as for the trench excavation: £1.44 × (4.51/4.04) = £1.607.

Example 20.2

Bill rates: disposal

(a) Deposit surplus spoil in heaps, wheeling not exceeding 100 m	m³	£1.10
(b) Deposit surplus spoil in heaps, wheeling not exceeding 150 m	m³	£1.36
(c) Spread and level surplus spoil, wheeling not exceeding 100 m	m³	£1.68

Variation account

Spread and level surplus spoil, wheeling not exceeding 200 m	m³

Item (b)	£1.36	
Less item (a)	1.10	
Wheeling 50 m	0.26	
Wheeling 100 m	0.52 plus item (c)	£2.20

In this example only part of the rate is proportioned – the wheeling. The cost of the wheeling is isolated, proportioned and the result used to adjust a bill rate, giving the analagous rate.

Example 20.3

Bill rate: concrete work

Concrete 1:3:6 in beds 100–150 mm thick, finished to receive quarry tile.	m²	£56.87

Variation account

Concrete 1:2:4 in beds 100–150 mm thick, finished to receive quarry tile.	m³

Labour will be the same for mixing and laying, So calculate material costs for both mixes as follows.

Cement	£52.95/tonne delivered to site
Sand	£14.17/tonne delivered to site
Aggregate	£12.02/tonne delivered to site

Mix 1:3:6

1 volume OPC	1.44 t/m³ + 2.5% waste	£78.15
1.25 hours/t unloading at £5.30/hour		9.78
3 volumes sand	1.52 t/m³ + 10% waste	71.08
6 volumes aggregate	1.6 t/m³ + 5% waste	121.16
		280.17
Shrinkage add 25%		70.04
9 m³ costs		350.21

1 m³ costs	35.02

Mix 1:2:4

1 volume OPC	1.44 t/m³ + 2.5% waste	78.15
1.25 hours/t unloading at £5.30/hour		9.78
2 volumes sand	1.52 t/m³ + 10% waste	47.38
4 volumes aggregate	1.6 t/m³ + 5% waste	80.77
		216.08
Shrinkage add 25%		54.02
6 m³ costs		270.10
1 m³ costs		45.02

Bill rate per m³	£56.87
Profit and oncost is 20%: so deduct £56.87/6	9.48
Net rate	47.39
Deduct cost of 1 m³ of material 1:3:6 mix	35.02
If arguments earlier are correct, plant and labour cost is	12.37
Add cost of 1 m³ of material 1:2:4 mix	45.02
	57.39
Add profit and oncost 20%	11.48
Variation account rate per m³	£68.87

Example 20.4

Bill rate: facing brickwork

One brick thick wall built with facing bricks (PC £245 per 1000) in English bond, weather jointed in cement/lime mortar 1:2:9.	m²	£35.00

Variation account

One brick wall built with facing bricks costing £326.85 per 1000 in Flemish bond, weather jointed in cement/lime mortar 1:2:9.	m²

Extra cost of bricks is £81.85 per 1000
English bond requires 89 bricks/m² faced on side
Flemish bond requires 79 bricks/m² faced one side
Additional material cost: £326.85 × (79/1000) − £245 × (89/1000) = £4.02 per m²

Bill rate	£35.00
Profit oncost 20%: so deduct £35.00/6	5.83
	29.17
Add material cost	4.02
	33.19
Profit and oncost 20%	6.64
Variation account rate per m²	£39.83

This calculation ignores the fact that for each square metre of wall there are ten fewer facing bricks but ten more common bricks to lay. It is considered that the additional cost of this labour is negligible.

Example 20.5

Bill rate: plaster work

2 cts Carlite plaster 13 mm thick to brick walls.	m^2	£5.57

Variation account

(a) 2 cts Carlite plaster 10 mm thick to block walls.	m^2	
(b) 2 cts do on ingoes not exceeding 300 mm wide.	m	

Materials
Carlite browning £120/tonne delivered to site, £126.63 unloaded
Carlite finish £95/tonne delivered to site £101.63 unloaded
Browning covers 140 m^2/tonne 11 mm thick, 160 m^2/tonne 8 mm thick
Finish covers 450 m^2/tonne 1.6 mm thick
Unloading 1.25 hours/tonne at £5.30

Browning 11 mm thick per m^2	£0.90
Browning 8 mm thick per m^2	0.79
Materials saving per m^2	0.11

Labour
Assume a squad of 2 craftsmen and 1 labourer.

Cost per hour: 2 at £6.10	£12.20
1 at £5.30	5.30
Squad cost	17.50

Squad will apply: 5.5 m^2/hour 10 mm thick in two coats
4.5 m^2/hour 13 mm thick in two coats

For work in narrow widths the output is halved.

Cost to apply two coats browning 13 mm thick per m^2	£3.89
Cost to apply two coats browning 10 mm thick per m^2	3.18
Labour saving per m^2	0.71

Variation account rate item (a)

Bill rate per m^2		£5.57
Profit and oncost 20%: so deduct £5.57/6		0.93
		4.64
Deduct: material saving	£0.11	
labour saving	£0.71	0.82
		3.84
Add profit and oncost 20%		0.77
Item (a) from variation account: rate per m^2		£4.61

Note that only the browning coat was adjusted and only the cost of the material.
There is no need to bother with the skim coat of finish.

Variation account rate item (*b*)

Bill rate per m^2	£5.57
Profit and oncost 20% so deduct £5.57/6	0.93
	4.64
Deduct labour cost for 4.5 m^2/squad hour at £17.50	3.89
	0.75
Add labour cost for 2.75 m^2/squad hour at £17.50	6.36
	7.11
Add profit and oncost 20%	1.42
Cost per m^2	£8.53

Item (b) from variation account, 300 mm wide
Rate per m   £2.56

21

Computer estimating systems

Introduction

In recent years the demand for computer aided estimating has increased, resulting in the development of a number of computer estimating systems. This chapter examines these systems and their characteristics.

The first section describes four major components of such systems:

1 The resources which form the basis of the estimate. A large part of this book deals with the build-up of resource rates.
2 The computer library structure.
3 The pricing medium. Very often this is a bill of quantities prepared by a quantity surveyor, but it could also be a rough take-off prepared by a contractor from plans and specifications.
4 The use of computer data. The information entered into the computer may be used for other purposes; for instance, the transfer of the net rate for an item of work to the costing system may permit an assessment of profitability.

The second section of the chapter describes the chronological sequence of operations undertaken by an estimator in pricing a plan and specification contract using a computer aided estimating system.

There follows a review of two popular estimating systems, chosen for their diverse characteristics. The chapter concludes by considering spreadsheet programs, which are particularly useful for estimating. Many of the examples in this book have been developed using spreadsheets, with information from one spreadsheet being transferred electronically to another.

Components of computer estimating systems

Resources

The majority of computer estimating systems hold labour, plant and material resources in a data bank which is normally termed the resource library. Labour and plant are held as rates per hour, which are updated as elements of the rates change. Material rates are held in the same format as that in which they are quoted, e.g. a price per 1000, and are normally updated at the time of pricing a new project (see Table 21.1).

In Chapter 2, on labour costs, an illustration was given of the build-up of the all-in hourly rate from its various components. In a computer system this would be an exercise carried out and maintained as the components of the rate changed, rather than for each contract. For example, if the holiday with pay stamp cost increased on 30 June, then the rate for labour in the computer system would also have to be changed on that date. Any estimate prepared subsequent to that date would automatically use the current rate in the system, which would incorporate the increased cost of the holiday with pay stamp.

Table 21.1   *Example resource library*

Code	Description	Unit	Rate
L100	Labourer	hours	
L110	Bricklayer	hours	
P100	Concrete mixer	hours	
M100	Common flettons	1000	
M110	Mortar	m^3	

Similarly, the rates for plant held by the computer system would be derived from build-ups similar to those detailed in Chapter 3. If the cost of fuel were to increase then this component of the plant rate would be changed and the rate for plant increased. If a rate for a hydraulic excavator is required, then the rate indicated by the computer would be based upon the latest rates for each component of the plant rate.

As mentioned above, labour and plant rates are normally expressed in terms of a cost per hour. When these rates are called up by an item description the work output constant will be applied, so that the element of labour and plant in each item description will be converted to a rate per unit of measure. This is described in the trade chapters in this book.

Material resources are normally held by the computer in the manner in which they are quoted; for example, bricks will be held as a cost per 1000. This allows the library to be updated as the material resource costs increase or decrease. Most computer systems update materials resource costs at the time of tender for each new contract. It would be normal to compile the tender using the rates held in the resource library, and at the same time to obtain quotations from suppliers for all material required for the particular project. As quotes are received the resource library would be updated and the project in hand automatically repriced.

Some computer systems allow resources to be held as composite items. For example, the item for a softwood door frame may include the costs of the material and of the labour and sundry materials in fitting and fixing. If the labour is referred to a labour resource item, then an increase in the holiday with pay stamp will also be recorded as an increase in the cost of the door frame.

Most computer systems allow profit and overheads to be added as a percentage, spread evenly over all of the items. Alternatively they may be added in differing amounts to individual items or groups of items, resulting in an uneven spread. The profit and overheads addition may be positive or negative, allowing the cash engineering of a tender.

Library structures

Two library structures have been adopted by suppliers of computer aided estimating systems.

First there are the printed libraries, often also sold as price books. These systems are characterized by coded printed lists of work descriptions, usually based upon the *Standard Method of Measurement*, which are rigidly structured and priced. To operate this system the estimator first calls up a description of work by code and enters the quantity of work. The item is then priced at the rates held in the library. Although rigidly structured, these systems allow for: items additional to those held in the library; the repricing of resources; the alteration of labour and plant constants and materials wastage factors; and the combination of items to

Table 21.2 *Example bill library*

| Code | Description | Unit | Resources | | | |
			Code	Unit	Quantity	Rate
F1104	102.5 mm bkwk	m^2	L100	hours	0.5	
			L110	hours	1.0	
			M100	1000	48	
			M110	m^3	0.1	
						£12.60

form composite items. This type of system requires a large amount of computer storage for the comprehensive price book; however, computer storage is inexpensive and access times are acceptably short.

The remaining systems allow the estimator to build up a library and a hierarchical structure. Such a system has the disadvantage of requiring a large investment of time in setting up, but once structured it will more likely reflect the working method of the estimator and will be more efficient in terms of the computer store required by the library.

Table 21.2 shows an example bill library.

Pricing medium

The pricing medium can be of two types: either a bill of quantities supplied by a quantity surveyor, or a list of quantities prepared by the estimator. A bill of quantities is a contract document and will have been prepared in accordance with the *Standard Method of Measurement*. The list of quantities prepared by the estimator is for the contractor's use only and may be as detailed as considered necessary by the estimator.

Most computer estimating systems are configured to allow the estimator to prepare his own measured quantities from drawings and specifications and subsequently to price them.

Measurement by the estimator

There are a variety of methods for the input of measurement by the estimator. For those systems based upon the use of a price book type of library, it is necessary to match the bill of quantities descriptions with the descriptions in the price book. The code of the item in the price book is entered, and this will call up a *Standard Method of Measurement* type of description.

For those systems based upon an estimator configured library, the estimator must first build a coded library of resources. When the library of resources is complete, the estimator will enter a description against which the codes and amounts of resources are attributed.

The quantities measured from the drawing are entered against the description, this being derived from a price book library or a library compiled by the estimator. The quantities are entered by means of a digitizer or an electronic ruler, or by typing in through the keyboard. Most systems will retain the quantities and permit them to be printed as a record.

Pricing a bill of quantities

For those systems without a *Standard Method of Measurement* structured library, the bill of quantities must be represented within the computer. In some systems this is done by typing the item reference from the bill of quantities followed by the unit quantity of the item. In other systems the whole bill of quantities is retyped into the computer. Advances in optical character readers (OCRs) have made this process less awesome; one supplier advertises that a 100 page bill of quantities can be read into the computer in one hour. Against each item of the bill of quantities is set the resource rates multiplied by the constant for the particular bill item, resulting in a unit quantity rate. The result of this process is a computer printed priced bill. It should be noted that the resources and constants must be entered for each bill item, unlike the price book method where the resources and constants are already set against each item. However, the priced bill of quantities, with its attendant resources and constants, may be stored as library data to be accessed at some future date.

Use of computer data

An advantage of holding data on a computer, as opposed to solely on paper as in the majority of cases currently, is that the data may be re-sorted and used in another form. The following is a description of how data held by a computer aided estimating system may be used to give useful information at a post-contract stage.

Breakdown of prices in the bill of quantities or estimator's schedule

It was stated above that unit rates are derived as follows:

unit rate = (labour rate × constant) + (plant rate × constant cost/unit) + material rate

The total for a bill of quantities item is the quantity of the item multiplied by the unit rate. A breakdown bill is often prepared by estimators by entering the labour, plant and material elements of each rate on the printed bill of quantities (Table 21.3). Most computer aided estimating systems will allow the production of the bill of quantities in unpriced, priced and breakdown bill formats.

Resources schedule

As stated above, the multiplication of the unit rate by the quantity of the item will give the total cost of the item. However, the computer can in addition multiply the quantity of the item by the labour constant. Then by summing all like labour it is possible to calculate the total amount, in hours, of each trade required for the contract. A similar calculation will give the total hours required of each type of plant and the total amount of each type of material. The computer aided estimating system can therefore produce a full resources schedule for the contract with no additional information input (Table 21.4). Attached to each item on the schedule can be the

Table 21.3 *Example of a breakdown bill*

Item	Quantity	Unit	Rate	Labour	Plant	Material	Sub-contractor S	Prime cost and provisional	Profit
22/C	10	m³	£24.20	£5.60	£1.50	£15.10	—	—	£2.00

Table 21.4   *Example of a resources schedule*

Resources		Total units		Total
Code	Description	required	Rate	net of profit
L100	Labourer	2500 hours		
L110	Bricklayer	4000 hours		

rate that was used in the estimate for the purchase of material. This information will be invaluable to a contractor's purchasing department.

Valuation options

A useful valuation option can be provided by the computer. First, all the quantities in the bill are set to zero. Then the quantities of work carried out to the date of the valuation are input. The computer aided estimating system multiplies the work carried out by the rates in the bill of quantities, giving the total value of work to date (Table 21.5). Some systems allow the input of a percentage of a particular item completed to date in lieu of an exact quantity.

For the client the valuation will be printed in a gross format, that is with the profit and overheads addition or deduction. In order that the contractor can compare the cost with the tender, some systems allow the printing of both a gross and a net valuation. A net valuation is one without any adjustments for profit and overheads. In cases where the contractor wishes to engineer his cash flow to maximum advantage, the net valuation option is of great assistance in subsequent profit/loss assessments.

For fluctuating contracts using the formula method of increased cost recovery, a further useful option is to allow the insertion of NEDO price adjustment categories against the bill of quantities items. With the further input of the NEDO indices the computer can be programmed to produce a formula price adjustment increased cost claim automatically.

Cash flow forecasting

At the post-contract stage most contractors create a contract programme; indeed, a contract programme is required by many standard forms of contract. At this stage it is useful with computer aided estimating systems to date each item or group of items in the bill of quantities or the estimator's schedule of quantities. If each item in the bill or schedule has a start date and a completion date then the system can be programmed to re-sort the priced bill or schedule into date order and thereby produce a cash flow forecast for the project (Table 21.6).

Table 21.5   *Example valuation report*

Bill item	Description	Unit	Rate	Quantity this month	Quantity to date	Total to date
28F	102.5 mm bkwk	m^2	£12.60	100	2000	£25,200

Table 21.6   *Dated resources report*

Bill item	Description	Quantity	Unit	Rate	Start date	Duration
28F	102.5 mm bkwk	4000	m^2	£12.60	5/4/90	12 weeks

The addition of a date also permits the production of a schedule of when materials are required on site. If the resources library contains the normal delivery period for the material, this schedule can be expanded into a materials purchasing schedule.

A computer comparison of labour resources with the timing of items in the bill of quantities permits a report on the labour requirements for the contract, and gives early warning of inefficient or impractical peaks in demand for one trade.

A continual computer monitoring system which compares the valuation value to date with the anticipated cash flow will also provide early warning of slippage in the contract programme or lack of productivity.

Links with the cost system

All companies are required by law to keep accounts of their business. Most contractors run a comprehensive accounting practice which allows an assessment of costs to date. Where these costs are recorded on a project level it is possible to merge the estimating system with the costing system in order to automatically produce a profit/loss account.

Suppliers of computer aided estimating systems approach this in two ways. The first is to code the accounts with the references of items in the estimate and transfer cost information to the estimating system. The second is to code the items in the estimate with a cost centre number from the accounts system and transfer estimating data to the accounts. Either way the valuation measure provides the multiplier for the assessment of net income. The comparison with the account of cost to the same date will give a measure of profit/loss.

Some computer aided estimating systems do not permit the electronic transfer of data between systems, but do allow the valuation to be coded with the account system cost centre codes.

Comparison and incorporation of suppliers' and subcontractors' quotes

A characteristic of quotations from suppliers is that they show a price for each item. Ironmongery quotations, for example, may have been obtained from four suppliers, all of whom have submitted a price list. It is not possible to judge which is the most advantageous until the prices have been multiplied by the quantities in the bill of quantities. Some computer aided estimating systems permit the insertion of a number of quotations in the resources file, and subsequently produce a printed report of all of the items using the ironmongery resources priced for each supplier. The most advantageous supplier's rates are chosen and the system prices all items at these rates, deleting the rates of the other suppliers.

The rates of several subcontractors can be inserted against the items in the bill of quantities and judged in a similar manner.

This facility is also useful if new quotations are sought at a post-contract stage.

Chronological analysis of computer aided estimating

1   The estimator receives an invitation to tender accompanied by plans and specifications and perhaps a bill of quantities. The first task for the estimator is to open a new job file on the computer with details of the proposed contract. When no bill of quantities is received the estimator has to prepare his own take-off. This is done in one of two ways:
    (a) The estimator measures from the drawings and prepares a builder's quantities bill on paper.
    (b) The estimator uses the facility included with some computer aided estimating systems for direct input to the screen. This requires that the estimator first calls up the appropriate library description, or codes and forms a library description. Either way, dimensions are set against the description and the computer carries out the necessary squaring and totals the number of units for each item description. These are stored in the computer memory either as dimensions or as totals. In the latter case the computer will provide a printed trace of all the dimensions and squaring.
         In some computer aided estimating systems the measurement can be carried out using an electronic or sonic digitizer, i.e. the measurements are transferred electronically from the drawing to the screen.
2   The estimator is now in possession of a take-off of all quantities, whether provided or not. Three situations can now exist:
    (a) In case 1(b), where the estimator measured direct into the computer against coded item descriptions, the bill of quantities will be resident in the computer's memory. In this instance the estimator will press the pricing button and the computer will price and total the bill automatically from the data in the resources library.
    (b) If the estimator has prepared the take-off on paper, he has now to identify the item descriptions in the library or prepare item descriptions for the library with their associated resources. Against these descriptions can be set the total number of units from the take-off. This being completed, the extensions and totalling can be carried out by the computer.
    (c) If the bill of quantities is supplied to the estimator then the item descriptions and totals are transferred from the bill to the computer. This is done
        (i) Either by identifying a description from the bill of quantities in the computer aided estimating system library and entering the code
        (ii) Or by typing in the item description from the bill of quantities. In this case resources from the resource library will be associated with the description.
3   Following these permutations of presentation and input, a situation now exists where all of the information is resident in the computer. The estimator must now check the pricing of his resources. Traditionally this was done by photocopying pages from the bill of quantities and sending these to the suppliers and subcontractors. Now most computer aided estimating systems produce a schedule of resources, giving the total number of hours of various classes of labour, the total number of hours of various classes of plant and the total quantities of materials. Depending on the system, items can be selected for pricing by subcontractors and these are printed as mini-bills. The materials schedules are sent to suppliers for pricing.
4   As quotations are received from subcontractors and suppliers, their rates are entered either in the resource library or against the item descriptions. An advantage of a computer system is that competitive quotations can be incorporated very late in the process.
5   A bill priced at net rates is presented to management for their addition of overheads and profit and any final adjustment. These adjustments are incorporated into the final tender bill.

6   A priced, printed bill of quantities is produced to accompany the tender.
7   If the tender is successful, the estimator will transfer a copy of the computer data to the surveying division for use in valuations, NEDO calculations, cash flow and profit/loss activities.

Review of two popular estimating systems

Wessex

Distributed by:
Wessex Electronic Publishing Ltd
Park Place, North Road, Poole, Dorset, BH14 0LE

The Wessex computer aided estimating system requires a microcomputer with at least a 20 megabyte hard disk. The system is built around the *Wessex Database – Building*, a comprehensive price book containing over 500,000 price elements.
   Contract information is input in one of two ways:

1   Information from a bill of quantities is coded according to the reference in the price book. The price book items are described in accordance with SMM6. Item descriptions not in the price book (called 'rogues') can be created and coded, after which associated resources are referenced.
2   Measurement from drawn information is input directly into the computer, first by entering the item description code from the price book, and secondly by entering the dimensions via the computer keyboard using a sonic digitizer.

Approximate quantities measurement is either by using the approximate estimating section in the price book, or by aggregating item descriptions to form a new coded item description. Up to ten levels of aggregation are possible.
   Any section of the measurement may be factored to allow for multiple units on a contract.
   All descriptions and resources represented in the price book may be modified and added to. Additions to the price book rates must be made for overheads and profit. These additions can be made per item, per trade, per labour type or by any other identifiable classification. The addition is as a percentage and may be positive or negative. The system, however, does not relate bill activities to time, and a cash flow forecast is not produced automatically.
   The output from the Wessex system includes:

● A priced bill of quantities in trade, element or NEDO category order. This can also be produced net or gross of percentage overheads and profit additions.
● A priced resources schedule, giving the total labour, plant and material requirements for the contract. This information can also be given for each individual trade.
● A materials listing from the unpriced resources schedule for material.
● A print of the original dimension input, since all the input information is retained.
● A number of reports such as on labour productivity.

The Wessex computer aided estimating system is aimed at the management of the estimate. It therefore has no electronic link into such post-contract activities as profit/loss accounting and order scheduling.

Manifest

Distributed by
Manifest
16a Molivers Lane, Bromham, Bedford, MK43 8JT

The Manifest computer aided estimating system requires a microcomputer and envisages the use of a hard disk machine. Disk space usage is around 6000 tender items per megabyte.
    Contract information is input in one of two ways:

1   From a bill of quantities, the bill item reference is entered with the unit and quantity. An abbreviated or full description can be entered as required by the user.
2   Measurement from drawn information is input directly into the computer, first by entering an item reference code and an abbreviated or full description, and secondly by entering the dimensions via the computer keyboard or using a Calcumator (an electronic scale with a keypad).

Once the bill entry is complete, each item is called up and the code for each resource is entered together with the quantity or constant of that resource. The resource library is common to all contracts and is compiled by the user. A ready made library can be supplied if required. Resources or activities can be aggregated to form a composite item, but there can be only one level of aggregation.
    Any section of the measurement may be factored to allow for multiple units on a contract.
    Additions to rates are made for overheads and profit. These additions can be made per item or per labour, plant, material, subcontractor, prime cost or provisional sum, or special category. The addition is as a lump sum on the items or as a percentage on the categories, and may be positive or negative. The system, however, does not relate bill activities to time, and a cash flow forecast is not produced automatically.
    The output from the Manifest system includes:

●   A priced bill of quantities in trade, element or NEDO category order. This can also be produced net or gross of percentage overheads and profit additions.
●   A priced resources schedule, giving the total labour, plant and material requirements for the contract. This information can also be given for each individual trade.
●   A materials listing from the unpriced resources schedule for material.
●   A print of the original dimension input, since all the input information is retained.
●   A number of reports such as on labour productivity.

Spreadsheets

A spreadsheet is an electronic worksheet held within a computer. It can be considered as a large sheet of paper ruled vertically and horizontally. The columns and rows are referenced; in this way every block or cell on the worksheet has a reference. The size of a cell is always one line, but it can vary in length (depending upon the program) from 1 to 80 characters. The size of the worksheet is dependent on the memory of the computer, but currently 1000 rows by 100 columns is not unusual. At any one time the computer screen displays a very small part of the whole worksheet, usually about 20 rows by 80 characters. If an oblong hole of about 120×80 mm were to be cut in a piece of A5 paper, and this paper were to be moved around a piece of A3 lined paper marked off in 15 mm columns, the effect would be the same as the amount of the worksheet which is at any one time visible on the computer screen.

The power of the spreadsheet comes from its ability to accept text, numbers or a formula into each referenced cell. The formula is compiled with reference to another cell, and hence columns of figures may be totalled and the total used in another cell as part of a multiplication formula.

Consider a very simple example. Assume that someone with an income of £800 per month wishes to determine whether it is possible to save £1500 for a holiday in June. A worksheet could be configured as in Table 21.7. Within the worksheet the income is entered as £800 in cell B4 (column B, row 4). Cells C4, D4, E4, F4 and G4 all refer to B4 and therefore all carry the value £800. Cell C3 refers to B12 and therefore dispays the same value. Cell D3, E3, F3 and G3 are referenced in a similar manner. Cell B12 contains the formula

Table 21.7 *Household accounts spreadsheet* (£)

	A	B	C	D	E	F	G
1		Jan	Feb	Mar	April	May	June
2							
3	Bt fwd		220.00	570.00	870.00	1040.00	1340.00
4	Income	800.00	800.00	800.00	800.00	800.00	800.00
5	Rent	150.00	150.00	150.00	150.00	150.00	150.00
6	Gas	80.00			80.00		
7	Elect.	50.00		50.00		50.00	
8	Food	200.00	200.00	200.00	200.00	200.00	200.00
9	Petrol	100.00	100.00	100.00	100.00	100.00	100.00
10	Car tax				100.00		
11							
12	Total	220.00	570.00	870.00	1040.00	1340.00	1690.00

Table 21.8 *Updated household accounts spreadsheet* (£)

	A	B	C	D	E	F	G
1		Jan	Feb	Mar	April	May	June
2							
3	Bt fwd		170.00	470.00	720.00	840.00	1090.00
4	Income	800.00	800.00	800.00	800.00	800.00	800.00
5	Rent	200.00	200.00	200.00	200.00	200.00	200.00
6	Gas	80.00			80.00		
7	Elect.	50.00		50.00		50.00	
8	Food	200.00	200.00	200.00	200.00	200.00	200.00
9	Petrol	100.00	100.00	100.00	100.00	100.00	100.00
10	Car tax				100.00		
11							
12	Total	170.00	470.00	720.00	840.00	1090.00	1390.00

B3 + B4 − B5 − B6 − B7 − B8 − B9 − B10. The other cells in row 12 are similarly referenced. Over £1500 is available by June for the holiday.

If the rent were to change to £200 per month, the value in cell B5 would be changed, and the sheet would recompute to that in Table 21.8 in less than 1 second. The holiday at £1500 is now not possible.

If it is considered how many A4 size forms can be held on a single worksheet, and then how they all interrelate, the power of the spreadsheet becomes apparent. Many spreadsheets contain sophisticated copying and editing facilities which makes compiling the sheets very easy. Many also allow data stored on a totally separate worksheet to be accessed by the worksheet resident in the computer's memory. After considering the above and the fact that all the examples in this book were made up on spreadsheets, it is logical to conclude that the spreadsheet is a very useful tool for the modern estimator.

22

Tender strategy

Introduction

A tender strategy is a plan made by a construction company in order to achieve growth, increased market share or other objectives as stated in the company plan. Construction companies are in business and therefore seek to gain wealth, which may be in cash or in assets such as buildings, land and plant.

It is generally perceived that in order to increase profit, turnover must also be increased. Conversely, a decrease in turnover will lead to a decrease in profits. Companies therefore generally plan for growth, with periods of stabilization in order to reinforce resources. The strategy may include the targeting of types of work or types of client, and may also involve marketing of the company by advertising or by making presentations to targeted clients.

Construction, however, is very different from manufacturing. If a manufacturer wishes to increase turnover, his plan will be based upon the development of a product which is seen as being desired by the majority of the population. A car manufacturer, for instance, might develop a new model which is fuel efficient, large, comfortable, powerful and available in many configurations. The costs will be calculated and the vehicle priced at or below the level of the competition. A particular market will be targeted, e.g. fleet owners, family motorists or the 20s age group. A massive advertising campaign will be aimed at the chosen market, with offers of test drives, free cars for a day, low finance and so on. Errors in estimating costs which result in low profits can be adjusted by price increases, or more likely through a mixture of model and specification revisions and price increases; for example, the 70i becomes the 70xi, which is £1000 more but has electric windows as standard.

A construction company wishing to increase turnover has basically to make the offer – the tender – attractive to the client and the client's professional advisers. The contractor's product is his construction service. The building has been designed by the architect; the client is interested in procuring the building as designed at the least cost. It follows, therefore, that the contractor wishing to increase turnover has to tender at a lower margin for profit than his competitors. Any profit on a fixed cost is profit; however, unlike the manufacturer, the contractor cannot adjust for estimating errors by price increases or revisions to the specification. The opportunity for a loss increases as the profit margin is reduced.

A construction company can operate in the short term without profit but cannot operate at all without cash. For this reason any plan for the company must include a cash flow forecast for the period of the plan. This forecast will be updated as tenders are won and costs are realized.

Within the company plan, and specifically within the compilation of the tender, account must be taken of the risk and uncertainty associated with particular projects.

It is the intention of this chapter to give a flavour of the above subjects, to allow an appreciation of the importance of accurate estimating in the company plan, and to discuss the strategies which can be devised in order to increase turnover and profitability.

The company plan

The construction company, unlike its manufacturing counterpart, can increase turnover without dramatically increasing its capital assets. A manufacturer looking to increase turnover by increasing the sales of a product must also increase production. This requires an increase in the production facility, usually a factory containing machinery owned by the manufacturer. This production facility requires capital. The construction company's production facility is the site, which is owned by the client. The construction company's manufacturing resource is construction plant, which is commonly hired.

A manufacturing company is a relatively stable environment with smooth cash flow; that is, income and costs are fairly static over a short period and flow in and out on a regular basis. It is rare for a manufacturer's liability to suppliers etc. at any given time to exceed the assets. For this reason manufacturing companies can raise capital by selling shares, which basically represent the value of the assets of the company.

By comparison, a construction company is an unstable environment. Cash flows on individual projects vary; normally the highest expenditures and incomes fall in the middle third of a project. For all but the largest companies it is common for total liabilities to be greater than total assets. For these reasons, construction companies find difficulty in raising capital for projects by selling shares and are forced to finance projects through borrowing. On the other hand, construction company profits are potentially high in relation to capital invested.

The following example illustrates the application of a tender strategy. A builder, a private limited company, has a turnover of £10 million per annum. The builder's work comprises light industrial buildings, local health authority health centres and ambulance stations, local authority housing, petrol station forecourts and buildings, and small, high quality housing developments for a local speculator. All work is obtained by competitive tender except for the private housing work, for which the tender is negotiated.

In the company plan the builder has listed the following:

Assets
- Office block of 500 m² with yard and joinery workshop of 200 m², valued at £750,000.
- Sundry small plant and a wheeled tractor excavator with loader and backhoe. Total value £40,000.

Strengths
- Knowledge of the local area.
- Five highly skilled joiners and one apprentice.
- Four dedicated contracts managers.
- Capital invested of £200,000.

Weaknesses
- No track record with projects of over £2 million or with projects of longer than 18 months.
- No reputation with national clients other than one petrol company.
- Very traditional mode of operation.

Objectives
- To increase turnover to £50 million per annum during a five year period.
- To increase turnover to £15 million per annum in the following year.
- To sell top quality service rather than reducing tender margins.

Methods
- To target national clients, particularly hotel chains and supermarkets whose requirements are related to shopfitting and joinery.
- To investigate and sell quality control.
- To investigate and sell value management.

- To investigate and sell construction management and management contracting.
- To prepare sales literature, visit clients and advertise locally.

Resources
- Employ two contracts managers with experience in management contracting.
- Employ quality controller.
- Employ planner skilled in use of Critical Path Method.
- Employ accountant to release director to new sales office.
- Employ one extra staff in sales, estimating/buying and accounts/wages.
- Build 300 m² extension to office.
- Reserve £20,000 for computers, fax and new telephone system.
- Negotiate bank overdraft facility of £1 million secured against office, workshop and yard.

Tender strategy
- Do not tender for design/build work at this juncture since this may prejudice commissions from local architects.
- Negotiate work on the basis of cost plus 5 per cent overheads and 4 per cent profit.
- In each month, tender for work on the basis of 5 per cent overheads and 2 per cent profit for the first £2.5 million, then 5 per cent overheads and 4 per cent profit for the remainder.
- If a tender is not accepted in first five attempts, reduce profit percentage addition to zero.
- Do not tender for work where the number of contractors tendering is greater than that recommended in the various codes of procedure for tendering.

In this example the builder is attempting to increase turnover through a variety of actions. To increase turnover in a construction market which is buoyant and expanding is easier than in a stable or a contracting market, since in the latter situation work is gained at the expense of competitors. To penetrate new markets or to increase market share within a sector requires extensive marketing and salesmanship.

It should be noted also that the plan contains a commitment to increase manpower resources, at current wage levels, worth approximately £350,000 per annum. This represents a 7 per cent overhead on the £5 million of increased turnover, and therefore in the first year of the new trading plan a small loss is anticipated on the overhead account. This will be corrected as turnover increases in the following year when no further expansion in staffing is anticipated.

The company plan therefore is a set of interactive objectives aimed at achieving greater wealth. In times of decreasing demand in the construction sector, the company plan will be a statement of the methods by which the contractor can retain resources in order to be in a good position to take on work when an upturn arises. This can result in the contractor planning to make a loss on selected projects, that is buying work.

Tender policy

The probability of winning

The probability of winning a tender is theoretically based on the level of mark-up for overheads and profit. If it is assumed that all estimators can accurately estimate the cost of carrying out work and that for a particular project all estimators plan for the same method of construction, then the only variable is the level of mark-up. It follows therefore that the winning tender will be that which has the lowest level of mark-up. This theoretical basis, however, does not perform well in a live situation since, as is discussed below, there are many factors which influence the final estimated figure.

One of the problems with competitive tendering is that the lowest tenderer generally wins. It is a fallacy therefore to think that by cutting margins on one estimate, subsequent estimates can be increased to cover the loss. Mark-ups for overheads and profit tend to move on an industry-wide basis, increasing when many projects are seeking contractors and decreasing when many contractors are seeking projects.

It follows that the successful contractor will be the one which observes the economic indicators and measures the likely future demand for work by reference to the overall performance of the economy. While this statement is generally true there is a problem in that the government, through its various agencies, is the construction industry's biggest customer – and the government tends to use the construction industry as a public spending regulator. The construction industry therefore may receive little warning of a sudden downturn in demand. Good construction industry forecasters and planners will be watching the economic indicators in the same way as the treasury, and will attempt to guess at the manner in which projects from government agencies are likely to be released.

Reasons for variability between estimators' prices

As stated above, if it is assumed that all estimators can accurately estimate the cost of carrying out work, and that for a particular project all estimators plan for the same method of construction, then the only variable is the level of mark-up. This is plainly false. There are many reasons why estimators' prices vary, and the purpose of this section is to consider those reasons.

Items of responsibility

In pricing a tender, the estimator will have regard not only to the plans and bills of quantities but also to the specifications, preliminaries and conditions of contract. From these documents the estimator will build up a feel for the project and will overlay his base prices with an addition for complying with quality control provisions, work carried out under abnormal conditions, and so on.

Quantities

In situations where a bill of quantities is provided, the quantities are fixed. However, where the estimator is responsible for preparing quantities there is obviously scope for variability in measurement.

Material costs

Suppliers will offer special discounts to contractors who purchase large quantities of material in a given year. Thus large national contractors can obtain high levels of discount, which is reflected in the pricing of items or in the contractor's margin. Additionally estimators will use different allowances for off-loading, handling, storage and wastage.

Method of construction

A major factor in the variability of estimators' prices for items is the method for the carrying out of the work. The method will be derived at a meeting of the estimator, the planner and the buyer, together with the contracts manager and contracts quantity surveyor who have a preliminary designation to the project. At this meeting the method of construction, extent of subcontracting, timing of work items, availability of material, availability of in-house plant and other resources, and so on, will be discussed and will be recorded by the estimator. The pricing will be based upon the agreed method.

To illustrate the variability in method, consider the number of ways in which a 150 mm *in situ* reinforced concrete suspended slab can be constructed in accordance with given drawings and specification. The concrete, however it is delivered to the project, can be placed in a number of ways. First, it may be barrowed, lifted by hoist, and barrowed again into position. Secondly, it may be lifted by tower crane and the tower crane skip emptied in the required position. Finally, it may be pumped to the required position. The method chosen by the team will depend on the relative costs of each. Remember that if a contractor has a tower crane in the plant yard which is unlikely to be working otherwise during the concreting periods then this could be the cheapest option, notwithstanding the fact that it may be the most expensive option on a hiring comparison.

Labour

Two variables are associated with labour, namely gross unit rate and productivity. Whereas the former can be relatively easily established (see Chapter 2), the productivity aspect is more difficult. Research into the productivity rates of labour has shown wide differences between projects and between estimators' assumptions. This can also be demonstrated by observing the differences between constants assumed in estimating books such as this.

Plant

As with labour, plant is subject to both cost and productivity variables.

Site conditions

Whilst current standard contract conditions largely negate this subject as an unknown risk item, variations between cost and estimate can arise if the exact significance of the information provided is not appreciated by the estimator. For instance, if the presence of running sand in a trial pit is not noted by the estimator a low estimate may be made in the excavation work.

An example of a condition which will affect the estimate but may not be contained in the contract documents is a restriction in access. If the only direct route to the site has a height, width or weight restriction then this may affect the transport of plant and/or material.

Location

Construction costs vary throughout the UK. Studies demonstrate that building in the north of Scotland is characteristically over 20 per cent more expensive than building in the south-west of England. Local builders will be aware of the costs in their particular location, but national builders will adjust their standard data based on the location of the contract.

Escalation factors

In contracts which state that the tender price is to remain firm, estimators must make an allowance for increasing costs during the period of the project.

Contract time

It is common for the contract period to be specified by the design team, and the builder will plan resource requirements to fit the period stated. There is a theory that the method of construction adopted by the contractor is largely determined by the design and the period stated. That is, where there is a full design and a fixed period, all contractors tendering will choose the same method of construction. However, where parts of the design are left to the contractor, i.e. by including a performance specification in the contract documents, and/or the time is left to be stated by the contractor, then variability in work method between builders tendering is likely to occur.

Overheads and profit

As discussed above, these are the traditionally perceived areas of greatest variability. Profit is normally added to the tender amount as a percentage based upon a tender strategy. The overhead is a percentage added to the tender sum to cover the cost of the head office and supervision on site. An approximation of the percentage can be derived by taking the value of overheads from the previous year's audited accounts and dividing this sum by the turnover.

Contingency

This is an addition to cover contingent and unforeseen costs. It is accepted by the builder that the estimator will not have foreseen every eventuality and therefore a sum should be added based upon experience or on a calculation of risk factors (see later). However, since any addition reduces competitiveness, the contingency addition has to be carefully considered.

Cash flow and financing

The subject of cash flow is considered later, but it should be mentioned here that every project requires some degree of funding. The builder's income will not match his costs plus overheads and profit, since payment by the client is normally made up to six weeks in arrears and then less a retention of up to 5 per cent. The builder will be required to finance this shortfall in funding, normally through a bank arrangement. The tender should therefore include the finance charges on the overdraft.

Errors

It is commonly stated that the builder who wins a contract through competitive tendering is the one who makes the largest error. Procedures are outlined in the various codes of procedure for tendering for situations in which mathematical errors occur. If the contractor discovers an error of any other nature, the only recourse is to withdraw the tender (the offer) before it is accepted by the client.

Cash flow

The cash flow of a builder is defined as the resultant of income and expenditure expressed as a function of time. Cash flow is considered positive when income exceeds expenditure and

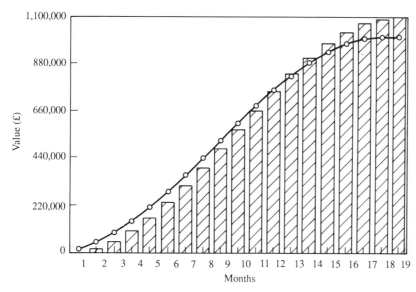

Figure 22.1 *Cash flow diagram: the curve represents expenditure, the bars income*

negative when expenditure exceeds income. The profile of a typical project (Figure 22.1) shows a negative cash flow for the first 12 months of an 18 month contract, i.e. the majority of the contract period.

The expenditure is normally shown as a curve, since costs arise on a weekly basis for labour and a monthly basis for subcontractors, although for logistical reasons the valuation and payment dates are spread through the month. Suppliers are normally paid at the end of the month following the month in which the material was delivered.

Income, on the other hand, is by interim payment which is made by the client on a monthly basis, normally two weeks following the end of the valuation month. This is represented on the diagram by the vertical bars.

As stated earlier in this chapter, the highly volatile construction market, the financial structure of builders and the irregular peaking of demands for finance make building companies very vulnerable to temporary shortages of cash. The problem facing companies is therefore one of convincing the financier, usually a bank, of the fact that the shortage is only temporary. A bank would prefer to see a cash flow forecast of the type illustrated for each project, and an aggregated forecast for the company as a whole. Since large national contractors have many projects at various stages of completion, their aggregated cash flow tends to a horizontal line, that is a constant demand for cash. Small and medium companies with a smaller number of projects tend to show wide fluctuations in the company demand for cash. This demonstrates the paradox that large national companies who have the resources to carry out a cash flow forecast have no need to make one, whereas small and medium companies have fewer resources but need to monitor cash flow very closely.

Construction of a cash flow diagram

A cash flow diagram for a tendered project is important, since it permits the amount of finance required for funding the project to be determined. At the tender stage it also allows the repricing of items to minimize the finance required. A cash flow forecast, as illustrated in Figure 22.1, may be constructed as follows.

Derive cost curve

There are four methods of deriving a cost curve, which are described below. In each method the cost curve represents the tender value of the project excluding head office overheads and profit.

Head office overheads are deducted from the project cash flow and added to the aggregate of all project cash flows, which forms the company cash flow. Overheads are normally a constant value per month; they are costs which are generally incurred irrespective of the value of work being carried out on the various project sites.

The methods of deriving a cost curve are as follows.

Pricing the planning bar chart

This is the most time consuming but most accurate method. The priced quantities, whether in a proper bill of quantities or the estimator's own quantities, are grouped into the same aggregated work items as are described on the planning Gantt chart or bar chart. The total value of each bar chart item is entered on the left hand side of the chart. The total value is then divided by the number of weeks shown for the bar chart item, and the resultant figure is entered in each week of the bar. The weekly cost for the project is then derived by adding vertically for each week.

An accumulative value, i.e. the total cost of work done to date, is entered in the row below the weekly totals. This accumulative cost is then plotted on a graph (value on the vertical axis, time on the horizontal axis), where it will form the traditional S shape when completed. Figure 22.2 shows part of this process.

Thirds and quarters method

This is the crudest, quickest and most unreliable method. The total net value of the project (less profit and overheads as above) is divided by four. On a value/time diagram as before, four points are marked. The first is at (0,0). The second is at one-third contract time and one-quarter project net value. The third is at two-thirds contract time and three-quarters project net value. The final point is at full contract time and project net value.

Figure 22.3 shows this method for the example of Figure 22.1, that is for a project total value of £1,100,000, a project net value of £1,000,000, and a contract period of 18 months.

Formula method

In this method the cost curve is derived by formula. The best known is the Hudson formula, developed for use in planning DHSS spending on capital projects:

$$y = S[x + Cx \wedge 2 - Cx - 1/K(6x \wedge 3 - 9x \wedge 2 + 3x)]$$

where y is the cumulative value of work; x is the month in which expenditure y occurs divided by P; P is the contract period in months; S is the contract value; and C and K are parameters based upon the contract value (see Table 22.1).

Pricing the planning bar chart						
Weeks		1	2	3	4	5
	Value					
Set up site	4,000	2,000	2,000			
Piling	20,000		5,000	5,000	5,000	5,000
Excavate trenches	16,000			4,000	4,000	4,000
Concrete foundations	24,000				6,000	6,000
Brick substructure	38,000					7,000
Subtotal		2,000	7,000	9,000	15,000	22,000
Accumulative		2,000	9,000	18,000	33,000	55,000

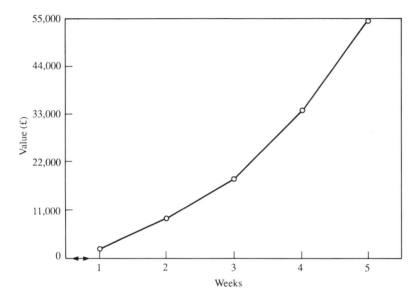

Figure 22.2   *Priced bar chart and S curve*

Table 22.1   *Parameters in Hudson formula*

	Parameters	
Contract value	C	K
£10,000 to £30,000	−0.409	7.018
£30,000 to £75,000	−0.360	5.000
£75,000 to £120,000	−0.240	4.932
£120,000 to £300,000	−0.200	4.058
£0.3 m to £1.2 m	−0.074	3.200
£1.2 m to £2.0 m	0.010	4.000
£2.0 m to £3.0 m	0.110	3.980
£3.0 m to £4.0 m	0.159	3.780

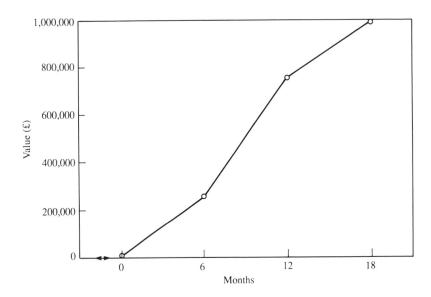

Figure 22.3   *The S curve by the thirds and quarters method*

Using the same example as above and the following BASIC program, 18 cumulative values are derived:

Project total value   £1,100,000
Project net value   £1,000,000
Contract period   18 months

```
10  FOR i=1 to 18
20  x=i/18
30  S=1000000
40  C=-0.074
50  K=3.200
60  y=S*(x+C*x∧2-C*x-((1/K)*(6*x∧3-9*x∧2+3*x)))
70  PRINT y
80  NEXT i
90  END
```

The derived values are shown in Table 22.2, and form the expenditure curve in Figure 22.1.

Balance sheet methods

More usually used for planning company cash flows as opposed to project cash flows, this method is based upon the assumption that the expenditure pattern for a company is repeated year after year. For any company the expenditure patterns are investigated from balance sheet records and plotted. The pattern is then plotted for the coming year based proportionately on the projected turnover. In this manner expenditure patterns for any cost centre on the balance sheet may also be examined, for instance head office overheads.

Table 22.2

Month	Value of y (£)
1	15,714
2	46,403
3	90,139
4	144,992
5	209,033
6	280,333
7	356,964
8	436,996
9	518,500
10	599,547
11	678,209
12	752,556
13	820,658
14	880,588
15	930,417
16	968,214
17	992,051
18	1,000,000

The above demonstrates the four main methods of deriving a forecast of expenditure. The first method – pricing the Gantt or bar chart – is the most accurate but manually is the most time consuming. The advent of computer aided estimating and planning has seen the development of computer packages which will produce a cash flow automatically from estimating and planning data. This is described in Chapter 21.

Insert income bars

Having derived the cost curve, the forecaster may now insert the income bars. The method involves first marking the cost curve at monthly intervals and reading off the monthly expenditure values. To these expenditure values is added the percentage value of overheads and profit, since the income received from the client will include these percentage additions. Payments from the client are normally received 14 days following the month in which the work was carried out, and therefore the income is plotted at the midpoint of each month as shown by the bars in Figure 22.1.

Review cash flow

The final operation in the cash flow analysis at tender stage is to review the financial implications of the forecast in terms of determining the amount of money which is required to fund the contract. As stated earlier, this is normally obtained from a bank and will be more easily acquired if it is supported by a forecast.

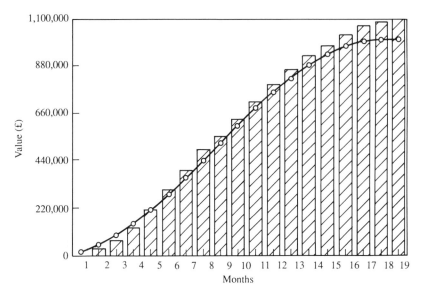

Figure 22.4   *The result of financial manipulation: the curve represents expenditure, the bars income*

The review stage permits the uneven allocation of the overheads and profit fund to minimize financing, therefore lowering the interest payable and reducing the tender. This operation is commonly called 'front loading' the tender. In extreme cases it is possible to artificially increase the amount in the overheads and profit fund by making negative overheads and profit additions to items of work which will be carried out later in the project. An example of a review to minimize financing is demonstrated in Figure 22.4 using the values obtained in the formula method example above.

Financial manipulation may also be used to artificially force a contract into loss in the earlier period where there is a requirement by the company to reduce trading profit in any particular trading year. For instance, if the current trading year is considered to be very profitable then tenders for projects in the remainder of the year can be 'back loaded'. That is, the overheads and profit will be attributed to those items carried out near the end of the contract, thereby transferring money from one trading year to the next. This has an advantage in that less corporation tax may be payable, but carries the risk that if the project is not completed for whatever reason then the builder will incur a loss. It also means that the money transferred is effectively lying in the client's bank account earning interest instead of in the builder's account. However, if the contract is one with a formula price adjustment clause then the loss of interest will be offset by a high increase in price recovery.

Risk and uncertainty

In a situation of risk, the result of an event is not known precisely but it is possible to determine:

● The number of possible results
● The probability of each possible result
● The value of the outcome of each possible result.

Uncertainty is defined as being a situation in which the determination of these factors is not possible. It is difficult therefore to calculate the effect of uncertainty upon the tender, and a discussion of uncertainty is out of place here.

A builder will see risks in two distinct categories:

● Those that can be insured against
● Those that cannot be insured against.

As far as the tender is concerned, risks such as fire, vandalism and theft will be insured against and the premium will be added to the tender in the preliminaries section. For those risks which may not be insured against, such as strike action, weather and failure in the power supply, an assessment will need to be made of the three factors above. Generally a good checklist of risks is given in the standard forms of contract, being those items for which the contractor is entitled to an extension of contract time but is prevented from claiming any loss and/or expense. In this situation the value of the risk item is the cost of maintaining the site with no work in progress.

The risk cost can be expressed as:

$$\text{risk cost} = \text{probability of event occurring} \times \text{period of event} \times \frac{\text{cost of delay}}{\text{unit of period}}$$

For example, if there is a 30 per cent probability that the site will be held up for one week due to bad weather, and the cost of maintaining the site for one week (including non-productive payments) is £5000,

$$\text{risk cost} = 0.30 \times 1 \text{ week} \times £5000 \text{ per week} = £1500$$

The £1500 will be added to the contract either in the preliminaries section or to the profit and overheads fund.

Some builders will have a checklist of risk items to analyse, although it is generally perceived that little work has been done in the direction of creating a data base for effective risk management.

Appendix

Cost of mortars, renders and fine aggregate screeds

Introduction

Plastic mixtures of cement and fine aggregates, or of cement, lime and fine aggregates, are used for a variety of purposes in the construction industry. The most important of these are:

- As mortars when building bricks and blocks
- As coatings on walls, ceilings and soffits, i.e. roughcasts and renders on a variety of bases such as brick, block, concrete, stone and lath
- As coatings on subfloors, i.e. screeds to receive floor finishes.

This appendix will deal with the calculations necessary to provide the cost of one cubic metre of mortar, render or screed ready for use on site. The term 'mortar' will be used for the remainder of the appendix.

The sample calculations have all been done using a computer and a spreadsheet. The computer is an IBM compatible PC and the spreadsheet program is called As-easy-as. This is a piece of shareware which produces files accessible by users of Lotus 123. A listing of the actual cell contents is included at the end of the appendix. Readers who are computer literate are recommended to type the listing into Lotus 123 or obtain a copy of As-easy-as. All major vendors of shareware and public domain software have it in their catalogue at a cost of from £3 to £10.

By adjusting basic costs and hours on a spreadsheet it is possible to obtain a thorough appreciation of the effects on the final cost of the material produced. Moreover, this can be done quickly, it can be updated as new raw materials and labour costs arise, and the results can be preserved for future use either as a computer file or on hard copy.

Data for mortar costing

The basic data required for mortar cost calculation are as follows.

Material costs and density constants

Materials are generally priced per tonne delivered to site. Constants of bulk density for aggregates vary according to moisture content and the mass of the aggregate material.

Shrinkage

The shrinkage of the total volume of dry materials to the final bulk of the wet mix is usually expressed as a percentage. It can vary depending on the type of aggregate used, the length of time for which mixing takes place, the addition of air entrainers etc. and the quantity of mixing water added.

Cost of labour

Small concrete mixers may require only one labourer to allow efficient operation. Larger machines require not only additional men but also some mechanical plant to load the drum efficiently. This appendix will confine itself to examples requiring up to two men, loading with hand tools.

Cost of plant

Most contractors own their own concrete mixer(s). However, use of these or hired equipment does represent a charge against a contract. How this charge is priced out is shown in Chapter 3. The hire rates used in this appendix have been extracted from trade sources current at the time of calculation. They represent only the 'capital' element of the cost of the machine; fuel, oil, grease, cleaning etc. are added as part of the running cost, i.e. the mixing operation. Maintenance of owned plant is treated as an overhead and is thus included in the oncost added to the final bill rate. Hired plant is maintained by the hire company, and this cost is reflected in the hire rate. Delivery to and uplift from site have not been included in the calculations.

Operational hours

While plant is charged to the site for each hour it is physically present on site, it does not always follow that it is actually productive for each of these hours. Generally, hire of concrete mixers is by the week. The drum seldom turns for 39 hours in any normal week. From his knowledge of the quantity of mortar required and the number of tradesmen applying it, the estimator will calculate how long the mixer must operate to provide that quantity.

For example – and a fairly simplistic one – suppose the estimator has a bill with one measured item: a one brick thick wall. The quantity is 2000 m^2, and he proposes to put a squad with eight tradesmen onto the bricklaying. Eight men will lay 400 to 500 bricks per hour, so the number of hours is $(2000 \times 118)/500 = 472$, or about 12 weeks. The brickwork requires $[(2000 \times 118)/1000] \times 0.6 = 142$ m^3 of mortar.

Using a 5/3.5 concrete mixer with an output of 3.5 ft^3 (0.102 m^3) of wet mortar per mix, the drum has to be loaded and emptied 1390 times in that 12 weeks. The output of a 5/3.5 concrete mixer is generally reckoned to be 1 m^3/hour, so 142 m^3 will take 142 hours. This is roughly 6 minutes per batch. A more optimistic output might be 5 minutes per batch of 0.102 m^3, giving a running time of 116 hours. Divide by 12 weeks and we have the number of 'operational hours' per week for the mixer – say 10 hours. So the hire cost is divided by 10 hours in this case when calculating the cost of mixing.

Of course the bricklayers might well break all records and finish the work in 9 weeks. In that case the mixer would be operational for 13 hours per week. Hire costs are saved, there is no saving in fuel or other consumables, and the bricklayers would want a good bonus or piecework rate.

If the reader has followed the argument so far, it will be obvious that a larger mixer would not have to run for so long – but it would probably make more mortar than the bricklayers could comfortably use before it took an initial set. Then again, a smaller mixer would run for longer and would produce smaller batches – but these might be so small that the bricklayers would keep running short of mortar, and that would hurt their bonus earning possibilities. So the estimator has to use not just cost estimating but also resource scheduling, so that plant is matched to the job and the labour force available.

Generally, hours running are more than this example shows. A figure of 20 to 35 hours per week is reasonable. Not that the machine is then producing just mortar for building; it may well be providing small quantities of concrete, screeding and rendering material.

Manual costing

A typical calculation would look as follows.

Example: cement mortar 1:3, machine mixed

Materials	
Ordinary Portland cement delivered to site per tonne	£52.95
Labour unloading 1.25 tonnes/hour at £5.30/hour	4.24
	57.19
1 m^3 OPC weighs 1.44 tonnes at £57.19/tonne	82.35
3 m^3 washed building sand weighs 4.56 tonnes at £15.24/tonne	69.49
	151.84
Shrinkage on mixing is 20%: so add 20/(100−20)=25%	37.96
Cost for 4 m^3	189.80
Therefore 1 m^3 of material costs	47.45

Labour and plant		
5/3.5 mixer hire rate per week	£22.00	
1 labourer in charge at (£5.30 + 15p) × 39 hours	212.55	
1 labourer in attendance at £5.30 × 39 hours	206.70	
Cost per week	441.25	
Machine is used for 30 hours per week, allowing 9 hours standing time. Therefore cost per hour is £441.25/30		14.71
Cleaning and maintaining 0.5 hours per day of 8 hours: per hour £5.30/16		0.33
Fuel: petrol 1.14 litres at £0.39		0.44
Oil and grease: say		0.05
		15.53
Output mixing mortar is 1.15 m^3 per hour: therefore cost of mixing 1 m^3 is £15.53/1.15		13.50
Cost 1 m^3 without profit and oncost		£60.95

This is a fairly simple if time consuming calculation if different mortar types have to be costed or different mixers and operational hours are tested.

Spreadsheet costing

A quick way to do this is to set up the calculation on a computer. There is no need to have purpose written software, and no need to be able to program. The costing uses a standard piece of office software called a spreadsheet, referred to earlier.

The computer printouts shown in the following are fairly self-explanatory. The initial blank spreadsheet or template was made up for use by students, and incorporates the following principles:

1 Information is laid out in screenfuls, i.e. discrete operations are confined to the data which can be displayed on one screen at a time.
2 The important parts of the sheet – the answer you want and the supporting statements – are on the first screen.
3 Principle 2 means that printing out the first screen is all that is required to keep a hard copy for the file.
4 Details of the calculations are put on later screens.

The examples now follow. All are based on 30 operational hours per week for the mixer.

Example 1

Figure A.1 shows the first screen for the costing of a cement mortar 1:3 mixed using a 5/3.5 concrete mixer without the addition of plasticizer or anti-freeze. Screens to the right and below are utilized for calculations and the storing of constants, costs etc. Paging down or right allows access to these additional screens. The reason for keeping this information separate is because it would be updated once for the job in hand and the various mortars, renders or screeds costed out before updating for the next job.

Figure A.1 is now taken line by line. The first four lines contain the title and some general explanation. Line 5 is blank. Line 6 lists the ingredients allowed for in the sheet, and the header 'cost per metre cube'. The actual proportions by volume of cement, sand and lime are entered on line 7, together with a 0 under anti-freeze and plasticizer to show that these are not used. The cost per m^3 appears under the header as the calculation proceeds. The correct amount is only displayed when all variables have been entered on the spreadsheet.

```
MORTAR - ANY MIX - ANY PURPOSE
Enter proportions for cement/lime/sand.  Enter a '1' for Anti-freeze
or for plasticizer.  Proportions are by volume.   Additives are
calculated from the proportion of cement.

OP Cement  Lime    Sand    A-frze    Plast        COST PER METRE CUBE
     1       Ø       3        Ø        Ø          ¦  59.62 ¦
...........................................................
The default shrinkage allowance is       20 %. or give an alternative
shrinkage allowance here>>>        Ø  Use this allnce>    20 %
...........................................................
                        Choose mixer & Fuel
                        by entering a '1'
Mixer Hire --- Cost ---   Petrol  Diesel     Output for mortar
    4/3     19.ØØ per week    Ø       Ø     Ø.95 m3 per hour      Ø
    5/3.5   22.ØØ per week    1       Ø     1.15 m3 per hour      1
    7/5     3Ø.ØØ per week    Ø       Ø     1.75 m3 per hour      Ø
                             1       Ø
...........................................................
To update the cost of materials, constants, consumables, etc. PgDn
```

Figure A.1   *Example 1, Screen 1*

```
MORTAR - ANY MIX - ANY PURPOSE
Enter proportions for cement/lime/sand. Enter a '1' for Anti-freeze
or for plasticizer. Proportions are by volume.  Additives are
calculated from the proportion of cement.

OP Cement Lime    Sand    A-frze   Plast        COST PER METRE CUBE
    1       2       9       0        0         ¦  54.93 ¦
...............................................................
The default shrinkage allowance is        25 %. or give an alternative
shrinkage allowance here>>>      30   Use this allnce>     30 %
...............................................................
                        Choose mixer & Fuel
                        by entering a '1'           Mixer  Petrol Diesel  Lub
                        Petrol Diesel               cost   Cost   Cost    Oil   Grease
Mixer Hire --- Cost ---                Output for mortar
   4/3    19.00 per week    0     0    0.95 m3 per hour   0   0.00   0.00   0.00  0.00  0.00
  5/3.5   22.00 per week    0     0    1.15 m3 per hour   0   0.00   0.00   0.00  0.00  0.00
   7/5    30.00 per week    0     1    1.75 m3 per hour   1  30.00   0.00   0.16  0.06  0.00
                           0     1                          30.00   0.00   4.86  1.80  0.12
...............................................................
To update the cost of materials, constants, consumables, etc. PgDn
```

```
Enter the costs of the materials and labour.  Unload     OPC        76.25        Mixer cost per week      30.00
OP Cement 52.95 per tonne   1.44 tonne/m3   1.25 tonnes/hr  Lime       60.00        Fuel cost per week        4.86
    Lime  50.00 per tonne   0.60 tonne/m3   1.25 tonnes/hr  Sand      208.48        Lub oil and grease/wk     1.92
Wshd Sand 15.24 per tonne   1.52 tonne/m3                   A-frze      0.00        Labour - 1 for 4/3        0.00
   A-frze  0.98 per litre 100.00 ltr/tonne (Febspeed)       Plast       0.00        2 men for others        337.35
    Plast  0.60 per litre   4.00 ltr/tonne (Febmix Admix)                                                -----------
 Gen. Lab  4.25 per hour                                    Unloading:                                     374.13
 Lab i.c.  4.40 per hour                                    OPC         4.90        Clean & maintain 1/2 hour
                                                            Lime        4.08          per day of gen labourer  10.63
Fuel cost:                                                          -----------                            -----------
  Petrol  0.40 per litre      Give operational              353.71        Cost per week          384.76
  Diesel  0.34 per litre      hours per week   >    30 <    Shrink    151.59                                -----------
                                                                    -----------               Cost per hour  12.83
Lubrication - costs                                                   505.30                              -----------
  Lub oil  1.20 per litre                                           -----------       Output:       1.75 m3/hr
   Grease  0.80 per kg.                                     For one
                                                              m3      42.11          Cost to mix 1 m3         7.33
The following fuel constants and lub oil/grease constants rarely change  ============                        ============
```

```
          petrol  diesel  lub oil grease
            p       d       l       g
   4/3    0.24    0.20    0.03    0.003   fo
  5/3.5   0.40    0.29    0.04    0.004   v
   7/5    0.55    0.47    0.05    0.005   se
                          END
```

Figure A.2 *The complete spreadsheet*

Lines 8 to 10 are used to define the shrinkage of the dry ingredients when mixed with water. It is generally held that mortars containing lime will shrink more than cement-only mortars. The spreadsheet as shown defaults to 20 per cent, unless a proportion of lime is entered when the shrinkage changes automatically to 25 per cent. The shrinkage used in the calculation is shown opposite the legend 'Use this allnce >'; here the value is 20 per cent for the cement mortar. The allowance can be set to any value that the estimator wishes by entering it after the legend 'or give an alternative shrinkage allowance here >>>'. The value opposite 'Use this allnce >' will change to any positive value entered rather than the default allowances.

Lines 11 to 18 are used to choose an appropriate mixer. A 1 is entered against the choice, taking care to zero any other earlier choice. Choosing *two* mixers would create an inflated cost. The spreadsheet as written for these examples does not give error messages in these instances; however, they can be written into the system if it is felt necessary. The choice in this example is the 5/3.5 mixer, and the 1 entered is echoed on line 18 under 'Petrol' and in the last column opposite 5/3.5. These echoes are used in the calculations to ensure that the appropriate costs for the hire and the fuel usage are properly picked out. Lines 19 and 20 are merely a prompt to go to the screen below to update costs for materials etc. The complete spreadsheet is shown in Figure A.2.

A cell contents listing is given at the end of the appendix. Typing these data into the cells given using either As-easy-as or Lotus 123 will give you a working spreadsheet. However, you should give names to some of the cells; a list of the names is provided with the cell contents listing.

Example 2

Figure A.3 shows the first screen for the costing of a cement/lime mortar 1:1:6 mixed in the smaller 4/3 mixer. The final cost per m^3 is much less than in Example 1 for the reason that only one labourer is required to operate a 4/3 mixer while the others require two men. This is a purely arbitrary decision taken when the spreadsheet was made up. The sheet could easily incorporate some form of entry to allow the estimator to define the labour required more precisely. Note that the shrinkage allowance has changed to 25 per cent because there is lime in the mix. This will increase the cost per m^3.

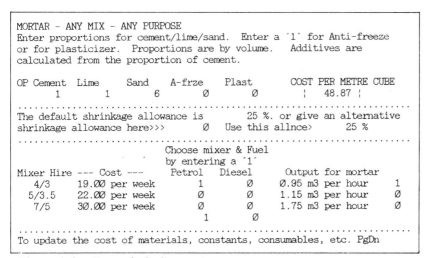

```
MORTAR - ANY MIX - ANY PURPOSE
Enter proportions for cement/lime/sand.  Enter a '1' for Anti-freeze
or for plasticizer.  Proportions are by volume.   Additives are
calculated from the proportion of cement.

OP Cement  Lime     Sand    A-frze    Plast       COST PER METRE CUBE
     1        1        6       Ø         Ø         ¦    48.87 ¦
..........................................................................
The default shrinkage allowance is        25 %. or give an alternative
shrinkage allowance here>>>      Ø   Use this allnce>      25 %
..........................................................................
                          Choose mixer & Fuel
                          by entering a '1'
Mixer Hire --- Cost ---    Petrol   Diesel     Output for mortar
   4/3     19.ØØ per week     1        Ø      Ø.95 m3 per hour    1
   5/3.5   22.ØØ per week     Ø        Ø      1.15 m3 per hour    Ø
   7/5     3Ø.ØØ per week     Ø        Ø      1.75 m3 per hour    Ø
                             1        Ø
..........................................................................
To update the cost of materials, constants, consumables, etc. PgDn
```

Figure A.3 *Example 2, Screen 1*

Example 3

Figure A.4 shows the first screen for the costing of a cement/lime mortar, mixed in the 4/3 mixer and with an imposed shrinkage of 30 per cent. The three volumes of binding agent (cement and lime) have the effect of increasing the cost per m^3; the higher proportion of sand lowers the cost per m^3, and the higher shrinkage allowance increases the cost per m^3.

```
MORTAR - ANY MIX - ANY PURPOSE
Enter proportions for cement/lime/sand.  Enter a '1' for Anti-freeze
or for plasticizer.  Proportions are by volume.   Additives are
calculated from the proportion of cement.

OP Cement  Lime    Sand    A-frze    Plast       COST PER METRE CUBE
     1       2       9        Ø         Ø         ¦   48.95 ¦
..................................................................
The default shrinkage allowance is        25 %. or give an alternative
shrinkage allowance here>>>       30   Use this allnce>      30 %
..................................................................
                             Choose mixer & Fuel
                             by entering a '1'
Mixer Hire --- Cost ---      Petrol  Diesel      Output for mortar
    4/3     19.00 per week      1       Ø       0.95 m3 per hour      1
    5/3.5   22.00 per week      Ø       Ø       1.15 m3 per hour      Ø
    7/5     30.00 per week      Ø       Ø       1.75 m3 per hour      Ø
                               1       Ø
..................................................................
To update the cost of materials, constants, consumables, etc. PgDn
```

Figure A.4   *Example 3, Screen 1*

Example 4

Figure A.5 shows the first screen for the same mortar as in Example 3 but mixed in the largest mixer, the 7/5. Note the increase in cost due to the doubling of the labour required to operate the machine as well as the increased running cost.

```
MORTAR - ANY MIX - ANY PURPOSE
Enter proportions for cement/lime/sand.  Enter a '1' for Anti-freeze
or for plasticizer.  Proportions are by volume.   Additives are
calculated from the proportion of cement.

OP Cement  Lime    Sand    A-frze    Plast       COST PER METRE CUBE
     1       2       9        Ø         Ø         ¦   54.93 ¦
..................................................................
The default shrinkage allowance is        25 %. or give an alternative
shrinkage allowance here>>>       30   Use this allnce>      30 %
..................................................................
                             Choose mixer & Fuel
                             by entering a '1'
Mixer Hire --- Cost ---      Petrol  Diesel      Output for mortar
    4/3     19.00 per week      Ø       Ø       0.95 m3 per hour      Ø
    5/3.5   22.00 per week      Ø       Ø       1.15 m3 per hour      Ø
    7/5     30.00 per week      Ø       1       1.75 m3 per hour      1
                               Ø       1
..................................................................
To update the cost of materials, constants, consumables, etc. PgDn
```

Figure A.5   *Example 4, Screen 1*

Example 5

Figure A.6 shows the first screen for the costing of a cement mortar with a plasticizer, mixed in the 7/5 mixer and with the default shrinkage allowance for cement mortars of 20 per cent.

```
MORTAR - ANY MIX - ANY PURPOSE
Enter proportions for cement/lime/sand.  Enter a '1' for Anti-freeze
or for plasticizer.  Proportions are by volume.  Additives are
calculated from the proportion of cement.

OP Cement  Lime    Sand   A-frze   Plast       COST PER METRE CUBE
     1        Ø       5      Ø       1         ¦   54.58 ¦
.............................................................
The default shrinkage allowance is      20 %. or give an alternative
shrinkage allowance here>>>       Ø   Use this allnce>     20 %
.............................................................
                        Choose mixer & Fuel
                        by entering a '1'
Mixer Hire --- Cost ---    Petrol  Diesel     Output for mortar
    4/3    19.00 per week    Ø       Ø      0.95 m3 per hour    Ø
    5/3.5  22.00 per week    Ø       Ø      1.15 m3 per hour    Ø
    7/5    30.00 per week    Ø       1      1.75 m3 per hour    1
                             Ø       1
.............................................................
To update the cost of materials, constants, consumables, etc. PgDn
```

Figure A.6   *Example 5, Screen 1*

Cell contents listing

The cell reference is given first followed by a colon, for example F8: or P50:. When entering the data, do not type in the colon. Some references are followed by information embedded in square brackets [ ]; this concerns column width or cell format. Do not type this in, but format the cell or set the width of the column as indicated; type in everything else. The apostrophe justifies text left, the caret ^ centres text and the quotes " justifies right. The @ is part of the function or operator in a formula. Be careful to distinguish between a zero and an upper case letter O. The former is not preceded by an apostrophe.

Table A.1 gives a list of cell names.

Table A.1   *Named cell list*

Name	Cell reference
AFR	B25
DFO	C43
DIES	B32
DSE	C47
DV	C45
GFO	E43
GRS	B36
GSE	E47
GV	E45
LFO	D43

Table A.1   *Named cell list (continued)*

Name	Cell reference
LIM	B7
LSE	D47
LV	D45
MATS	J37
MIXING	P37
OIL	B35
OPC	A7
OPH	G32
PETROL	B31
PFO	B43
PLA	E7
PSE	B47
PV	B45
SAN	C7
SHA	G10

A1: ´MORTAR – ANY MIX – ANY PURPOSE
A2: ´Enter proportions for cement/lime/sand. Enter a ´1´ for Anti-freeze
A3: ´or for plasticizer. Proportions are by volume. Additives are
A4: ´calculated from the proportion of cement.
A6: ^OP Cement
B6: ^Lime
C6: ^Sand
D6: ^A-frze
E6: ^Plast
F6: ´   COST PER METRE CUBE
A7: 1
B7: 2
C7: 9
D7: Ø
E7: Ø
F7: "¦
G7 [C2]: +MATS+Q34
H7: ´¦
A8: ´.........
B8: ´.........
C8: ´.........
D8: ´.........
E8: ´.........
F8: ´.........
G8: ´.........
H8: ´.........
A9: ´The default shrinkage allowance is
E9: @IF(LIM=Ø,20,25)
F9: ´%. or give an alternative
A10: ´shrinkage allowance here>>>
D10: 30
E10: ´ Use this allnce>
G10: @IF(D10=Ø,E9,D10)
H10: ´%

A11: ´.........
B11: ´.........
C11: ´.........
D11: ´.........
E11: ´.........
F11: ´.........
G11: ´.........
H11: ´.........
D12: ´Choose mixer & Fuel
D13: ´by entering a ´1´
J13: ^Mixer
K13 [W12]: ^Petrol
L13: ^Diesel
M13 [W5]: ^Lub
A14: ´Mixer Hire --- Cost ---
D14: ^Petrol
E14: ^Diesel
F14: ´   Output for mortar
J14: ^cost
K14 [W12]: ^Cost
L14: ^Cost
M14 [W5]: ^Oil
N14 [W6]: ´Grease
A15: ^4/3
B15 [C2]: 19
C15: ´per week
D15: Ø
E15: Ø
F15 [F2]: Ø.95
G15: ´m3 per hour
H15: +D15+E15
J15 [F2]: +B15*H15
K15 [F2 W12]: +PETROL*PFO*D15
L15 [F2]: +DIES*DFO*E15
M15 [F2 W5]: +OIL*LFO*H15
N15 [F2 W6]: +GRS*GFO*H15

A16: ^5/3.5
B16 [C2]: 22
C16: 'per week
D16: Ø
E16: Ø
F16 [F2]: 1.15
G16: 'm3 per hour
H16: +D16+E16
J16 [F2]: +B16*H16
K16 [F2 W12]: +PETROL*PV*D16
L16 [F2]: +DIES*DV*E16
M16 [F2 W5]: +OIL*LV*H16
N16 [F2 W6]: +GRS*GV*H16
O16 [S2]:
A17: ^7/5
B17 [C2]: 30
C17: 'per week
D17: Ø
E17: 1
F17 [F2]: 1.75
G17: 'm3 per hour
H17: +D17+E17
J17 [F2]: +B17*H17
K17 [F2 W12]: +PETROL*PSE*D17
L17 [F2]: +DIES*DSE*E17
M17 [F2 W5]: +OIL*LSE*H17
N17 [F2 W6]: +GRS*GSE*H17
D18: +D15+D16+D17
E18: +E15+E16+E17
H18: '
J18 [F2]: @SUM(J15..J17)
K18 [F2 W12]: (@SUM(K15..K17))*OPH
L18 [F2]: (@SUM(L15..L17))*OPH
M18 [F2 W5]: (@SUM(M15..M17))*OPH
N18 [F2 W6]: (@SUM(N15..N17))*OPH
A19: '........
B19: '........
C19: '........
D19: '........
E19: '..............
G19: '........
H19: '........
A20: 'To update the cost of materials, constants,
consumables, etc. PgDn
A22: 'Enter the costs of the materials and labour.
F22: ^Unload
J22: 'OPC
K22 [C2 W12]: +OPC*B23*D23
N22 [W6]: 'Mixer cost per week
Q22 [C2 W12]: +J18
A23: 'OP Cement
B23 [C2]: 52.95
C23: 'per tonne
D23 [F2]: 1.44
E23: 'tonne/m3
F23 [F2]: 1.25
G23: 'tonnes/hr
J23: 'Lime
K23 [C2 W12]: +LIM*B24*D24
N23 [W6]: 'Fuel cost per week
Q23 [C2 W12]: +K18+L18
A24: "Lime
B24 [C2]: 50
C24: 'per tonne
D24 [F2]: 0.6
E24: 'tonne/m3
F24 [F2]: 1.25
G24: 'tonnes/hr
J24: 'Sand
K24 [C2 W12]: +SAN*B25*D25

N24 [W6]: 'Lub oil and grease/wk
Q24 [C2 W12]: +M18+N18
A25: 'Wshd Sand
B25 [C2]: 15.24
C25: 'per tonne
D25 [F2]: 1.52
E25: 'tonne/m3
J25: 'A-frze
K25 [C2 W12]: +D7*AFR*D26*OPC*D23
N25 [W6]: 'Labour - 1 for 4/3
Q25 [C2 W12]: +H15*B29*39
A26: "A-frze
B26 [C2]: 0.98
C26: 'per litre
D26 [F2]: 100
E26: 'ltr/tonne (Febspeed)
J26: 'Plast
K26 [C2 W12]: +PLA*B27*D27*OPC*D23
N26 [W6]: ' 2 men for others
Q26 [C2 W12]: (H17+H16)*((B28+B29)*39)
A27: "Plast
B27 [C2]: 0.6
C27: 'per litre
D27 [F2]: 4
E27: 'ltr/tonne (Febmix Admix)
K27 [C2 W12]:
Q27 [C2 W12]: '-----------
A28: "Gen. Lab
B28 [C2]: 4.25
C28: 'per hour
J28: 'Unloading:
K28 [C2 W12]:
Q28 [C2 W12]: @SUM(Q22..Q26)
A29: 'Lab i.c.
B29 [C2]: +B28+0.15
C29: 'per hour
J29: 'OPC
K29 [C2 W12]: ((OPC*D23)/F23)*B28
N29 [W6]: 'Clean & maintain 1/2 hour
Q29 [C2 W12]:
J30: 'Lime
K30 [C2 W12]: ((LIM*D24)/F24)*B28
N30 [W6]: 'per day of gen labourer
Q30 [C2 W12]: +B28*5*0.5
A31: 'Fuel cost:
K31 [C2 W12]: '-----------
Q31 [C2 W12]: '-----------
A32: "Petrol
B32 [C2]: 0.395
C32: 'per litre
E32: 'Give operational
K32 [C2 W12]: @SUM(K22..K30)
O32: 'Cost per week
Q32 [C2 W12]: +Q28+Q30
A33: "Diesel
B33 [C2]: 0.345
C33: 'per litre
E33: 'hours per week   >
G33: 30
H33: '<
J33: 'Shrink
K33 [C2 W12]: +K32*(SHA/(100-SHA))
Q33 [C2 W12]: '-----------
K34 [C2 W12]: '-----------
O34: 'Cost per hour
Q34 [C2 W12]: +Q32/OPH
A35: 'Lubrication - costs
K35 [C2 W12]: +K32+K33
Q35 [C2 W12]: '-----------
A36: "Lub oil

```
B36 [C2]: 1.2
C36: 'per litre
K36 [W12]: '--------------
M36 [W5]: 'Output:
O36: (H15*F15)+(H16*F16)+(H17*F17)
P36: 'm3/hr
A37: "Grease
B37 [C2]: 0.8
C37: 'per kg.
J37: 'For one
J38: ^m3
K38 [C2 W12]: +K35/(OPC+LIM+SAN)
N38 [W6]: 'Cost to mix 1 m3
Q38 [C2 W12]: +Q34/O36
A39: 'The following fuel  constants and lub oil/grease
constants rarely change
K39 [W12]: '============
Q39 [W12]: '============
B41: ^petrol
C41: ^diesel
D41: ^lub oil
E41: ^grease
B42: ^p
C42: ^d
```

```
D42: ^l
E42: ^g
A44: ^4/3
B44 [F2]: 0.24
C44 [F2]: 0.2
D44 [F2]: 0.03
E44 [F3]: 0.003
F44: ^fo
A45: ^5/3.5
B45 [F2]: 0.4
C45 [F2]: 0.29
D45 [F2]: 0.04
E45 [F3]: 0.004
F45: ^v
A46: ^7/5
B46 [F2]: 0.55
C46 [F2]: 0.47
D46 [F2]: 0.05
E46 [F3]: 0.005
F46: ^se
D47: ^END
```

End of cell contents listing

Index